JN449290

테크는수다 in English

공대생을 위한 전공영어

손병희 · 강효선 지음

INTRODUCTION TO THE INTERNET OF THINGS

교우사

KYOWOOSA

머리글 PREFACE

우리가 살아가는 21세기는 지식과 정보가 새롭게 업그레이드되는 주기가 너무 빠르다. 불과 몇 달 전 신기술이었던 정보가 어느 새 구시대의 유물이 되곤 한다. 이러한 초 스피드시대에는 know-how나 know-where보다 언제 어디서나 가장 최근 정보를 액세스할 수 있는 능력이 있어야 경쟁에서 살아남을 수 있다. 이러한 경쟁력을 갖기 위해서는 유비쿼터스 기술과 영어능력이 무엇보다 중요하다.

본 교재는 이러한 시대의 흐름에 부합하여 유비쿼터스 기술에 대한 학습과 모든 지식의 매개체가 되는 영어에 대한 교육을 동시에 수행하기 위해 집필되었다.

본 교재의 구성은 다음과 같다.

먼저, 영문법에서 가장 중요한 단문을 중심으로 명사, 한정사, 대명사, 동사의 의미를 간단하게 Preview한 후 유비쿼터스 기술의 이해로 넘어간다.

유비쿼터스 컴퓨팅의 개념을 시작으로 유비쿼터스 기술의 기본 기술인 기술의 융합, RFID, 무선 통신인 센서 네트워크와 Ad-hoc 네트워크의 개념을 정의한 후 그 개념 속에서 유용하게 사용되는 영어 표현과 문법적인 요소들을 알아 본다.

그런 다음 유비쿼터스 기술의 포커스를 자동차로 옮겨 텔레매틱스 기술을 이해하고, 역시 유비쿼터스에서 빠질 수 없는 핵심 기술인 내장형 시스템, 즉 Embedded system에 대해 알아본다. 그리고, 유비쿼터스 기술의 마지막 기술이지만, 아무리 강조해도 지나치지 않는 보안 이슈에 대해 다룰 것이다.

교재 후반부에는 프레젠테이션에서 자주 사용되는 영어 표현을 익히고, 직접 영문 발표 자료를 만들어 영어 발표 능력을 배양할 수 있도록 'English for presentations' 세션을 준비해 보았다.

각 세션 마지막 부분에는 연습문제를 통해 유비쿼터스 기술과 영어를 다시 한번 정리한다. 미래를 준비하는 IT종사자와 더 나은 미래를 건설하기 위해 IT를 연구해야 하는 후학들에게 신기술의 트렌드와 영어라는 두 마리 토끼를 잡을 수 있기를 기대하며, 책을 내도록 아낌없이 지원해 주신 교우사 사장님, 관계자 여러분께 감사드린다.

이 책에서 다룬 내용이 현 시대에 비춰 다소 진부할지라도 급속도로 발전하는 기술의 영향력이라고 이해하고 읽어 주기를 바라며 아무쪼록 본 교재가 유비쿼터스 시대를 살아가는데 필요한 경쟁력을 기르는데 조금이나마 도움이 되길 바란다.

저자 **손병희, 강효선**

차례 CONTENTS

4 RFID

5 WSN

6 Ad-hoc Networks

7 Telematics

차례 CONTENTS

Preview

꼭 알아야 할 문법 구조

▸ **단문 (주절)**

– 명사
– 한정사
– 대명사
– 동사

여기서 단문을 더 구체화시키기 위해 형용사와 부사를 가지고 구체화 시킨다. 명사를 구체화 하는 것은 형용사이고, 명사가 아닌 것을 자세히 표현하기 위해서는 부사를 사용하게 된다. 한정사에는 지시형용사, 소유형용사, 그리고, 관사 등이 있다.

▸ **중문**

주절 + 접속사 + 주절
중문에는 등위접속사를 사용하게 된다. 등위 접속사에는 and, but, or가 있다.

▸ **복문**

주절 + 접속사 + 종속절
복문에는 명사절, 형용사절, 부사절로 구분이 된다. 이것은 종속 접속사가 이끄는 종속절의 품사와 연결된다.
명사절인 경우 that, if, whether, 그리고 w 의문사를 종속 접속사로 사용한다.
형용사절인 경우는 관계 대명사와 관계 부사를 가져온다.
마지막으로 부사절은 시간, 장소, 방법을 나타내는 종속 접속사를 이용해서 종속절을 만들게 된다.

이 교재는 영어 문법 교재가 아니므로 단문을 중심으로 꼭 알아야 할 문법요소를 다시 생각해 보는 시간을 갖도록 하겠다.

명 사

먼저 명사가 모든 언어에서 중요하다는 것을 인식해야 할 것이다.
왜냐하면, 문장 형식을 품사로 표현해서 보면 그것은 쉽게 알 수 있다.

품사로 표현한 문장 형식은 다음과 같다.

1. 1형식 (주어 (명사)+완전자동사)
   - I got up early this morning. (나는 오늘 아침에 일찍 일어났다.)
   - My heart leaps up when I behold a rainbow in the sky. (내 가슴은 뛴다/하늘의 무지개를 보면)
   - He speaks like a poet. (그는 시인처럼 말한다)
   - Actors appeared on the stage. (배우들이 무대 위에 나타났다)
   - There was a hole in his glove. (그의 장갑에 구멍이 하나 나 있었다)

2. 2형식 (주어 (명사)+불완전자동사+보어 (명사 or 형용사))
   - Roses smell sweet. (장미는 달콤한 냄새가 난다)
   - He seems kind to every girl. (그는 모든 여자에게 친절한 것 같다)
   - His face turned pale when everything was revealed. (그의 얼굴은 창백해졌다/모든 것이 드러났을 때)
   - Beauty is an abstract concept. (미는 추상적인 개념이다)

3. 3형식 (주어 (명사)+완전타동사+목적어 (명사))
   - He tied the kite to a boat. (그는 그 연을 배에 묶었다)
   - She bought three loaves of bread. (그녀는 빵을 세 덩어리 샀다)
   - We heard the sound of owls. (우리는 올빼미소리를 들었다)
   - He entertained us with his jokes. (그는 농담으로 우리를 즐겁게 해주었다)

4. 4형식 (주어 (명사)+수여동사+간접목적어 (명사)+직접목적어 (명사))
   - Cows provide us milk. (소는 우리에게 우유를 제공한다)
   - Mom gave me my allowance yesterday. (엄마는 어제 내게 용돈을 주셨다)
   - I will send you a copy of the report. (내가 그 보고서 사본을 내일 네게 보내줄게)
   - I owe Bill ten dollars. (나는 빌에게 십달러 빚지고 있다)

5. 5형식 (주어 (명사)+불완전타동사+목적어(명사)+목적보어(명사))
- They called **her** "Snow White." (그들은 그녀를 "백설공주" 라고 불렀다)
- I watched **something** float down the river.
  (나는 뭔가가 강물에 떠내려 가는 것을 보았다)
- Solar energy doesn't make **the air** bad.
  (태양에너지는 공기를 나쁘게 만들지 않는다)
- Please have **it** wrapped. (그것을 싸주세요)
- You shouldn't let **him** go there again.
  (넌 다시는 그가 그곳에 가게 허락하지 않는 게 좋겠다)

위에서와 같이 문장을 구성하는 구성요소는 대부분이 명사임을 알 수 있다.

이 때, 명사는 특이하게도 셀 수는 명사도 있고, 셀 수 없는 명사도 있다. 대부분 명사는 셀 수 있어서 절대적으로 셀 수 없는 명사는 따로 암기를 해야 한다.

셀 수 없는 명사는 다음과 같다.

1. 비슷한 물품으로 된 전체 그룹
   - baggage, clothing, equipment, food, fruit, furniture, garbage, hardware, jewelry, junk, luggage, machinery, mail, makeup, money, cash, change (잔돈), postage, scenery, traffic

2. 액체
   - water, coffee, tea, milk, oil, soup, gasoline, blood

3. 고체
   - ice, bread, butter, cheese, meat, gold, iron, silver, glass, paper, wood, wool, cotton

4. 기체
   - steam, air, oxygen, nitrogen, smoke, smog, pollution

5. 알갱이들
   - rice, chalk, corn, dirt, dust, flour, grass, hair, pepper, salt, sand, sugar, wheat

6. 추상적인 것들

- beauty, confidence, courage, education, enjoyment, fun, happiness, health, help, honesty, importance, intelligence, advice, information, news, proof, evidence, work, grammar, slang, vocabulary

7. 언어

- Chinese, English, Spanish

8. 학과목

- engineering, mathematics, psychology

9. 게임

- baseball, soccer, tennis, chess, poker, darts, billiards

10. 활동

- driving, studying, swimming, traveling, walking

11. 자연 현상

- weather, dew, fog, hail, heat, humidity, lightning, rain, sleet, snow, thunder, wind, darkness, light, sunshine, electricity, fire, gravity

모든 명사들은 적절한 한정사와 결합하여 명사구를 만들며, 가산명사와 불가산명사로 구분하여 단수인지 복수인지를 파악하여 명사와 대명사를 적절한 수로 일치시키는 것이 가장 중요한 포인트가 되겠다.

한정사

한정사는 관사, 소유 형용사, 지시 형용사, 수량 형용사, 전치 한정사 5가지로 구분해 볼 수 있다.

1. 관사

단수 명사에 대해서만 a/an을 쓴다. 그러나 the는 어떤 단어에나 쓰일 수 있다. 복수 명사에 대해서는 some을 쓴다.

the라는 정관사는 말하는 대상이 분명할 때 사용한다.

- 대표적으로 반복할 때
- 하나만 있을 때
- 명사 뒤의 구나 절이 어떤 대상을 말하는지 분명히 해 줄 때, 수식을 받아 무엇인지 분명할 때

2. 소유형용사

- my, your, his, her

3. 지시 형용사

- that, this

4. 수량 형용사

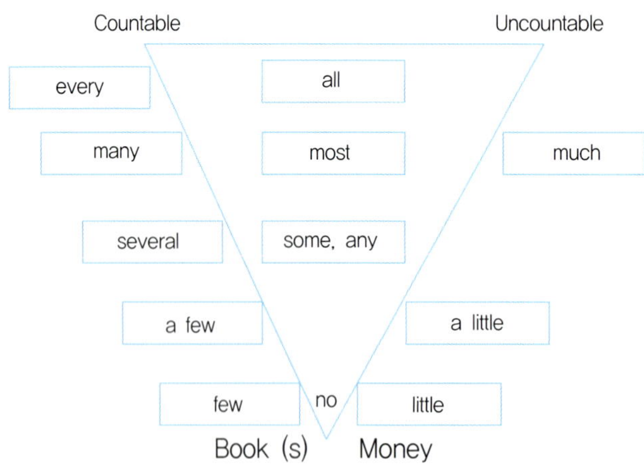

5. 전치 한정사

- all, so, such, 배수사

대명사

대명사는 인칭 대명사, 부정 대명사, 지시 대명사, it으로 구분한다.
대명사는 한번 나온 명사를 대신해 주는 명사이다.

1. 인칭대명사

- 주격, 목적격, 소유격, 재귀대명사

2. 부정 대명사

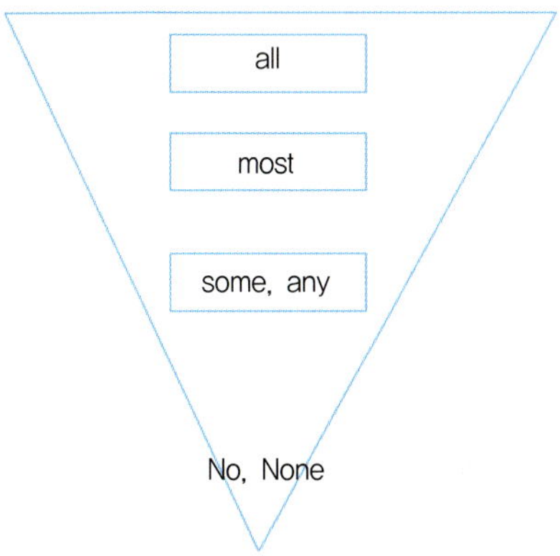

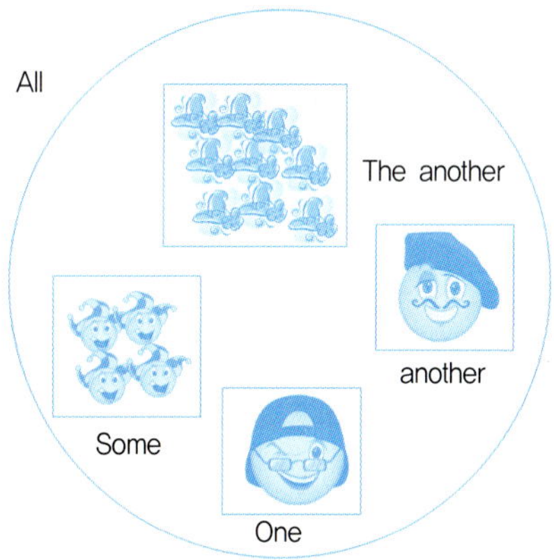

3. 지시 대명사

- The weather of Seoul is similar to that of Tokyo.

동사

12시제를 기억하자.

Tense	Aspect	단순 상태	진행	완료	완료 진행
Factual	현재	Complete 기정 사실	Incomplete	Priority 경험	진행 + 완료
Remoteness More polite Improbable	과거			~하자마자 ~했다	
	미래				

1

Ubiquitous Computing

1 Ubiquitous Computing

1.1 Introduction

Ubiquitous computing (ubicomp) is a **post-desktop** model of human-computer interaction in which information processing has been thoroughly **integrated into** everyday objects and activities. In the course of **ordinary** activities, someone "using" ubiquitous computing engages many computational **devices** and systems **simultaneously,** and may not necessarily even be aware that they are doing so. This model is **advancement** from the desktop paradigm.

This paradigm is also described as **pervasive** computing, **ambient** intelligence, or more recently, everyware.[1][2] When primarily concerning the objects involved, it is also physical computing, the Internet of Things, **haptic** computing,[3] and things that think.

The word, Ubiquitous computing, is suggested for the first time by Dr. Mark Weiser worked for PARC: Palo Alto Research Center in 1988

- From Wikipedia, the free encyclopedia

이번 장에서 꼭 알아야 할 단어

be integrated into

*If you **are integrated into** something, that means you have seamlessly become a part of something.*

- The town's modern architecture is very well **integrated with** the old.
- After brainstorming, the ideas you think of can **be integrated into** one concept.

ordinary

~으로 통합되다

*Something that is **ordinary** is normal or commonplace. It is not unusual.*

- She was a girl of **ordinary** beauty; she didn't stand out in a crowd.
- It is very difficult for an **ordinary** person to become rich and famous.

device

장치, 기구

*A **device** is something that is contrived or made.*

- A medical **device** is a product which is used for medical purposes in patients, in diagnosis, therapy or surgery.
- Our store sales numerous wireless networking **device** for home or office use.

advancement

발전, 진보

***Advancement** is a promotion or elevation to a higher rank or position.*

- There is little chance of **advancement** in my current job.
- She belongs to a national association for the **advancement** of Southeast Asians in Korea.

simultaneously

동시에

*If you do two things **simultaneously**, you do those things at the same time.*

- It is hard to pat your head and rub your stomach simultaneously.
- She was driving, texting, and smoking **simultaneously** when she got into a wreck.

pervasive

만연한,
구석구석 스며있는

*If something is **pervasive**, it is present or felt throughout place or thing.*

- There was a **pervasive** kimchi odor in the refrigerator.
- Mobile technologies are becoming **pervasive** even in many third world countries.

ambient

주위[주변]의

*Something that is **ambient** exists or is present on all sides.*

- I couldn't sleep in because of the **ambient** light coming in from the window.
  ☞ *Ambient music is a musical genre that focuses largely on the timbral characteristics of sounds, often organized or performed to evoke an "atmospheric", "visual" or "unobtrusive" quality. [공공장소에서 제공되는 조용하고 명상적인 느낌의 환경 음악의 한 장르]*

haptic

*Something that is **haptic** is related to or based on the sense of touch.*

☞ *그리스어로 '만지는'이라는 뜻의 형용사 'haptesthai'에서 온 말로 컴퓨터의 기능 가운데 사용자의 입력 장치인 키보드와 마우스, 조이스틱, 터치스크린 등을 통해 사람의 촉각을 느끼게 하는 기술이다.*

- **Haptic** technology interfaces with the user through the sense of touch.
- Samsung's **Haptic** phone is quite popular these days.

collocation이란 우리말로 연어(연결된 단어)라고 부른다. 예를 들어, 'ordinary'라는 단어는 'people'이라는 단어와 함께 자주 사용되어 '평범한 사람들'이라는 표현이 된다. 이 때 ordinary의 연어가 바로 people인 것이다.

Common Collocations

ordinary	device	advancement
Ordinary people Ordinary clothing Ordinary time	Manager device Audio device Input device	Advancement opportunities Supporting advancement Job advancement
pervasive	**ambient**	**haptic**
Pervasive technology Pervasive computing Pervasive language	Ambient temperature Ambient music Ambient light	Haptic feedback Haptic phone Haptic device

유비쿼터스 기술을 이해하는 데 도움이 되는 팁

- 유비쿼터스 컴퓨팅 vs. 가상현실

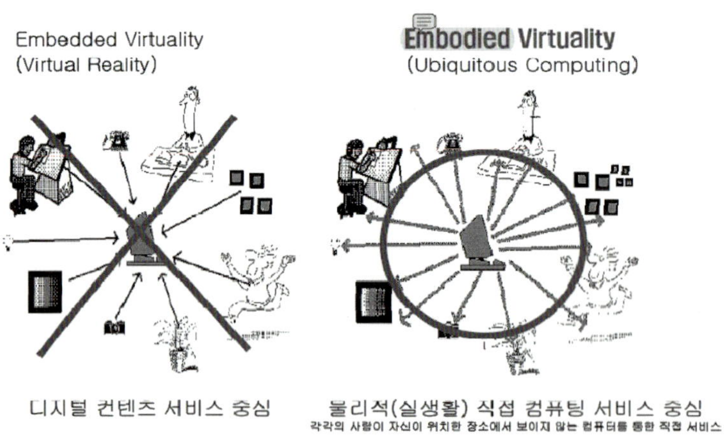

[그림 1] 유비쿼터스 컴퓨팅과 가상현실의 차이

(출처 : ITCT광주정보문화산업진흥원)

마크 와이저는 유비쿼터스 컴퓨팅은 가상현실(VR: Virtual Reality)과 거의 반대되는 개념이라고 하였다. 위의 그림에서 보는 것처럼 가상현실은 사람들을 컴퓨터로 구현한 세계 속으로 밀어 넣지만, 유비쿼터스 컴퓨팅은 사람들이 사는 세상 속에서 묻어나도록 만들어준다. 이처럼 유비쿼터스 컴퓨팅은 인간공학과 컴퓨터과학, 엔지니어링, 사회과학의 매우 어려운 결합이 될 것이라고 예견하였다.

유비쿼터스 기술을 이해하는 데 도움이 되는 팁

■ 유비쿼터스 컴퓨팅의 또 다른 이름

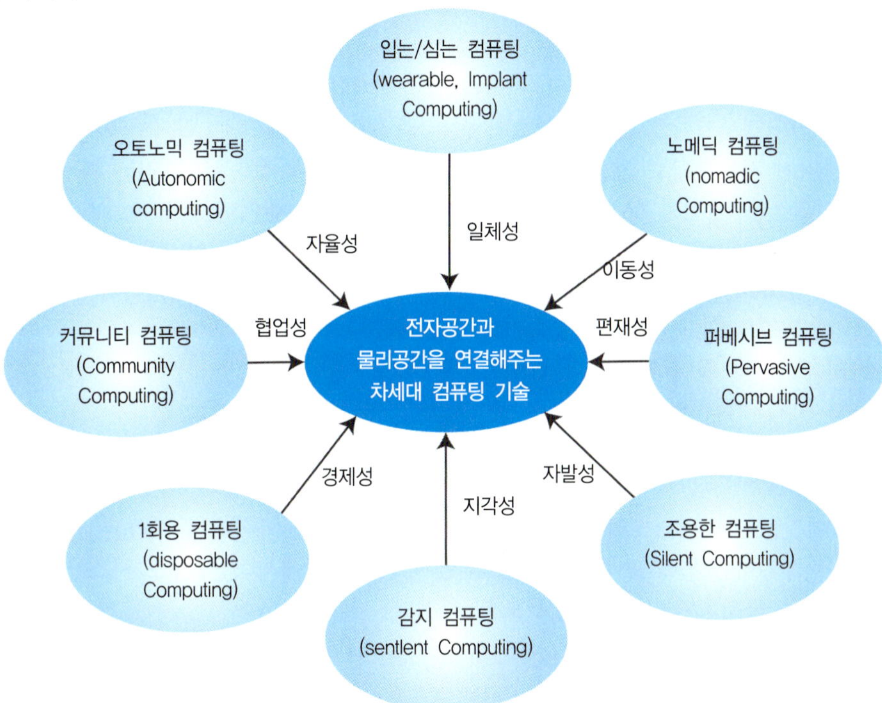

[그림 2] 유비쿼터스 컴퓨팅의 유사개념

(출처 : 다가오는 유비쿼터스 사회, 조위덕, 아주대 강의 2006년 4월)

- **퍼베이시브 컴퓨팅(Pervasive Computing)** : pervasive [pərvéisiv] a.퍼지는, 널리 미치는 스며드는. "널리 퍼지고 스며드는 컴퓨팅" 정도로 해석하면 될 것이다. 이 의미는 IBM에서 제안한 개념으로 "어디에나 있는" 뜻의 유비쿼터스 컴퓨팅과도 사전적 의미가 비슷하다고 할 수 있겠다. IBM이 생각하는 퍼베이시브 컴퓨팅은 첫째로 자동차와 핸들헬드 기기 등 각종 기기 안에 통합된 솔루션이며, 둘째가 무선영역으로 확장하기 위한 미들웨어이며, 마지막이 하드웨어와 소프트웨어, 서비스 등을 교차, 융합시키면서 새로운 비즈니스 모델을

창조하는 것이라고 한다.

- 조용한 컴퓨팅(Silent Computing) : 정보기기나 일상 사물에 심어진 컴퓨터들이 사용자가 의식하지 않더라도 사용자의 요구에 의해 일을 수행하는 컴퓨팅을 말한다. 이것은 마크 와이저가 발표한 "The Coming Age of Calm Technology"라는 논문에서 주장한 조용한 기술 (Calm Technology)이라는 개념에서 기반하였다.
- 두루누리 : 2004년 10월 19일 국립국어연구원에서 개설, 운영하고 있는 '모두가 함께하는 우리말 다듬기' 사이트를 통하여 '유비쿼터스(ubiquitous)'를 '두루누리'로 순화한 순수 우리말 신조어이다. 명사형으로 두루접속, 바로접속, 홍길동접속 등을 후보로 네트즌들의 투표를 통해 최다 득표로 선정되었다.
- 노매딕 컴퓨팅(Nomadic Computing) : no·mad·ic〔noumǽdik〕 a. 유목의, 방랑의 디지털 노매드 현상에서 유래한 용어로서 어떠한 장소에나 다양한 정보기기가 편재되어 있어 사용자가 굳이 정보기기를 휴대할 필요가 없는 환경을 뜻한다. 네트워크의 이동성을 극대화하여 사용자가 자유자재로 이동하면서 어디서든지 컴퓨터에 접속할 수 있는 환경인 것이다. 이른바 어디서든 연결된 (Always Connected)환경의 실현이다.
- 임베디드 컴퓨팅(Embedded Computing) : em·bed〔imbéd, em-〕

  vt. (embed·ded;embed·ding)

  1 [보통 수동형으로] 〈물건을〉 깊숙이 박다, 파묻다

  2 (마음 속 등에) 깊이 간직하다 《in》

  3 【언어·수학】 끼워 넣다

  vi. 끼워 넣어지다 《in》

  컴퓨터가 수행 기능을 미리 프로그래밍하여 사물이나 기기에 심어 지능화 시키는 기술을 말한다. 이것은 유비쿼터스라는 용어로 대체할 수 있는 개념이라기 보다는 유비쿼터스의 구성요건 중의 하나라고 볼 수 있는데, 후반 부에 이 책에서 임베디드 시스템, 임베디드 소프트웨어로 구분하여 자세히 다루도록 하겠다.
- 감지 컴퓨팅(Sentient Computing) : sen·tient〔sénʃənt〕a.

  1 감각[지각]력이 있는

  2 의식하는, 민감한

  센서를 통해 사용자의 상황을 인식하여 사용자가 필요한 정보를 제때 제공해 주는 기술을 말한다. 이러한 감지 컴퓨팅을 통하여 유비쿼터스 서비스를 제공할 수 있다. 이런 기술을 이용하여 u-커뮤니케이션 서비스, u-정보제공 서비스, u-상황인지 서비스, u-지능형 서비스 등을 받을 수 있다.
- 일회용 컴퓨팅(Disposable Computing) : dis·pos·a·ble〔dispóuzəbl〕a.

  1 처분할 수 있는, 마음대로 쓸 수 있는;〈소득 등이〉 (세금을 지불하고) 자유로이 쓸 수 있는

  2 〈종이 제품 등이〉 사용 후 버릴 수 있는

disposable diapers[cups, towels] 1회용 종이 기저귀[컵, 수건]
— n. [종종 pl.] 《미》 사용 후 버리는 물건, 일회용 물품
휴대폰 단말기 한대가 백만 원을 육박하는 이 시점에서 유비쿼터스가 일상화되기 위해 가장 선행되어야 할 기술이라고 생각되는 기술이다. 이 기술은 저렴한 가격의 컴퓨터로 1회용 종이처럼 가격이 매우 저렴하여 어떤 물건이라도 컴퓨터 기술을 활용할 수 있는 기술을 말한다.

- 입는 컴퓨팅 (Wearable Computing) : 컴퓨터를 안경이나 옷처럼 착용할 수 있게 해주는 기술이다. 컴퓨터를 인간의 몸의 일부로 만들 수도 있다. 이 기술은 앞으로 나노 기술, 바이오 테크놀로지의 힘을 입어 체내 이식형 컴퓨팅으로 발전해 나갈 것이다.
- 엔조틱 컴퓨팅 (Exotic Computing) : ex·ot·ic〔igzátik | –zɔ́t–〕〔Gk 「외국의」의 뜻에서〕a.
  1. 이국적인, 이국 정서의 ; 색다른, 낭만적인
     an exotic hairstyle 별난 헤어스타일
  2. 〈동식물 등이〉 외국산의, 외래의(foreign) 《종종 열대산을 말함 ; opp. indigenous》
     exotic plants 외래 식물
  3. 신종의, 실험적인
     exotic weapons 신형 무기

  – n. 외래 식물 ; 이국적인 취미, 외래어

  현실세계와 가상 세계인 전자공간과 물리공간을 연계해 주는 컴퓨팅 기술이다.

1.2 Useful Expressions

be aware of + 명사구 / be aware + that 주어 + 동사 : ~을 알고 있다	
• She is aware of his intention. • I am aware that it is not easy. • I was not aware that you were coming to the party. • She did not become aware of the problem.	• 그녀는 그의 의도를 알고 있다. • 나는 그것이 쉽지 않다는 것을 알고 있다. • 나는 네가 파티에 올 것이라는 것을 알지 못했었다. • 그녀는 그 문제에 대해 알지 못했었다.
be described as: ~이라고 설명되다, 묘사되다	
• The man was described by police as a terrorist. • I wouldn't describe her as a beautiful woman, but I would say she is pretty. • My husband was described by his friend as a noble person.	• 그 남자는 경찰에 의해 테러리스트로 묘사되었다. • 나는 그녀를 아름다운 여성이라고 묘사하지는 않을 것이지만 예쁘다고는 말할 수 있을 것이다. • 내 남편은 그의 친구들로부터 고상한 사람이라고 일컬어졌다.

1.3 Descriptive Grammar

information processing ***has been*** thoroughly ***integrated*** into everyday objects...

현재완료는 과거에 시작된 일이 현재까지 이어진 경우에 쓰이며, 형태는 have(has) + p.p(과거완료)를 사용한다. 예를 들어, "Everything is going well. We haven't had any problems so far."라는 문장을 보자. "모든 일이 잘 진행되고 있다. 우리는 (예전부터) 지금까지 어떤 문제도 없었다."는 의미가 된다. 과거부터 지금까지 계속 문제가 없었다는 정보는 알 수 있는 것이다. 본문에 나온 *'has been* thoroughly *integrated…'* 구문은 현재완료에 수동태가 결합된 구문이다.
수동태는 주어가 어떤 동작을 주체적으로 행한 것이 아니라 동작에 영향을 받는 것으로 be(get/become/grow) + p.p로 표현된다. "I was given a car by my brother." 이 문장에서 동사 'give'는 원래 '~에게 -을 주다'라는 이지만 'was given' 즉, 수동태로 사용되어 주어인 'I'가 누군가에게 무엇을 주었다는 의미가 아니라 받았다는 의미가 된다.
위의 문장을 다시 살펴보면,

information processing / has been (thoroughly) integrated / into everyday object
정보처리가 / (예전[과거]부터 지금[현재]까지 (철저하게) 통합되고 있다 / 모든 사물에…

영작해보기

S + have/has + been + p.p
예전[과거]부터 지금[현재]까지 ~되고 있다

① 그 계획은 예전부터 지금까지 다양한 방법으로 연구되고 있다.

→ ______________________________

[The plan / study / in various ways]

② 기술은 과거부터 지금까지 계속 빠르게 변화되었다.

→ ______________________________

[Technology / change / rapidly]

③ 그녀는 백신의 효과에 대해 계속 관심을 가지고 있다.

→ ______________________________

[She / research / effects of vaccines]

정답_ ① The plan has been studied in various ways.
② Technology has been changed rapidly.
② She has been researched the effects of vaccines.

1.4 Substance

마크 와이저 박사는 유비쿼터스 컴퓨팅이 메인프레임, PC에 이은 제3의 정보혁명의 물결을 이끌 것이라고 주장하였다. 첫 번째 물결은 대형 컴퓨터를 여러 명이 사용하는 메인프레임 시대로 하나의 컴퓨터를 여러 사람이 공유하는 환경으로 표현했다. 두 번째 물결은 1인 1PC를 갖는 퍼스널 컴퓨터시대를 지칭했다. 이 물결은 현재 우리들이 경험하고 있는 단계로서 개인적 컴퓨팅 영역 (Personal Computing Era)이라고 한다. 이 물결을 거쳐 1명이 여러 컴퓨터를 사용하는 유비쿼터스 컴퓨팅 시대가 2005년 이후에 일반화할 것으로 추정하였다. 마지막 물결인 3번째 물결은 이제 막 태동을 하려고 하는 유비쿼터스 컴퓨팅이며, 다른 말로 조용한 기술의 시대 (The age of calm technology)라고 말한다.

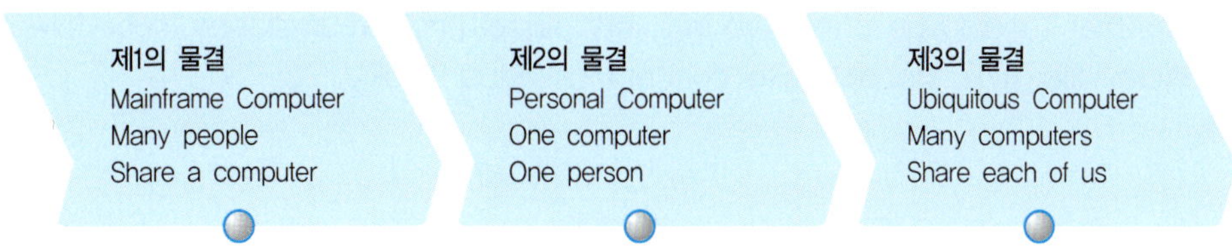

[그림 3] 컴퓨팅 패러다임의 진화

이 유비쿼터스 컴퓨팅은 다음과 같은 4가지 조건을 갖는다.

- 첫째, '연결되어야 한다. (Connected)' 이는 모든 컴퓨터와 사물 및 인간들이 서로 네트워크를 통해 연결되어야 한다는 것이다. 이것은 네트워크 연결의 5대 any화 (5 any 1: Anytime, Anywhere, Any network, Any device, Any service)[1]를 지향하고 있다.
- 둘째, '보이지 않는다. (Invisible; Disappearing)' 이는 수많은 컴퓨터와 디바이스 등이 흩어져 있지만, 사용자가 의식하지 못하도록 사용자의 눈에 보이지 않아야 한다는 뜻이다. 따라서, 일상 생활에 묻어나야 한다는 것이다.
- 셋째, '조용한 서비스라고도 한다. (Calm Service)' 이는 언제 어디서나 사용 가능해야 한다. 평소에는 보이지 않지만, 필요할 때는 사용자의 요구에 의해 사용할 수 있는 사용자 중심의 환경이라는 뜻이다.
- 넷째, '실제적이다. (Real)' 유비쿼터스 컴퓨팅의 기본 개념처럼 현실 세계에 존재하는 것이다. 이것은 모든 사물과 환경 속으로 스며들어 일상생활과 통합된다는 뜻으로 가상현실과는 반대의 개념이다.

영어 작문 연습

Ubiquitous Computing requires the following 4 conditions.
First, 'Connected.' It says every computer, object, and man must be connected with each through a network. This aims for 5 any – converted of network connection (5 any 1: Anytime, Anywhere, Any network, Any device, Any service)
Second, 'Invisible; Disappearing.' This means many dispersed computers and devices cannot be visible in order not to be cognized by users. So to speak, it has to be involved in everyday life.
Third, 'Calm Service.' It refers to being available anytime and anywhere. It is invisible during ordinary times, but being human-centered and ambient, it can be used in accordance with user's requirement.
Fourth, 'Real.' It exists in the real world like the basic concept of Ubiquitous computing. This means that it would be pervasive in every object and circumstance to be integrated into everyday life. It is the opposite concept of virtual reality.

1 5대 C: Computing, Communication, Connectivity, Contents, Calm

1.5 Glossary

- Ambient intelligence
- Context-aware pervasive systems
- Human-centered computing
- Human-computer interaction
- Sentient computing
- Ubiquitous services
- Smart devices
- Virtual reality
- Wearable computers
- Pervasive adaptation

References [참고문헌]

[1] 융합 그리고 넥스트, 손병희 저, 교우사, 2015

[2] Greenfield, Adam (2006). Everyware: the dawning age of ubiquitous computing. New Riders. pp. p.11-12. ISBN 0321384016.

[3] Hansmann, Uwe (2003). Pervasive Computing: The Mobile World. Springer. ISBN 3540002189.

[4] "World Haptics Conferences". Haptics Technical Committee, http://www.worldhaptics.org/hapticConferences.htm. Retrieved on 2007-10-13.

[5] Cambridge Advanced Learner's Dictionary, Cambridge University Press

[6] Collins COBUILD Intermediate Learner's Dictionary, Heinle & Heinle Publishers

[7] Longman Essential Activator, Longman National Corpus

[8] English Grammar in Use, Raymond Murphy, Cambridge University Press

1과 연습문제

01 Match each word in the first column with its definition in the second column.

1.	advancement	A.	at the same time
2.	simultaneously	B.	difuses through every part of something
3.	pervasive	C.	promotion or elevation to a higher position
4.	ambient	D.	based on the sense of touch
5.	haptic	E.	present on all sides
6.	ordinary	F.	become a part of something.
7.	be integrated into	G.	know about something
8.	be aware of	H.	normal or commonplace

02 Each of the following excerpts is about ubiquitous computing. Read the sentences. Fill in the blanks using the words in the below box.

1) ______________ is not a specific technology, but a scenario in which computers become more numerous and fade into the background, providing ______________ to human users and ______________ intelligence and computing capabilities in seemingly everyday objects.

> ubiquitous computing, embedding, information

2) ______________- almost imperceptible, but ______________ around us- is rapidly becoming a reality. How will it change us? How can we shape its ______________ ?

> everywhere, ubiquitous computing, emergence

3) Ubiquitous computing (ubicomp) integrates ________________ into the environment, rather than having computers which are distinct objects. Promoters of this idea hope that embedding computation into the _________ and everyday objects would enable people to interact with information-processing devices more naturally and casually than they currently do, and in ways that suit whatever location or ________ they find themselves in.

computation, context, environment

4) Ubiquitous computing and _________________ computing have been posed as polar _________________ even though they are often applied in very similar _________________.

applications, opposites, wearable

5) _________________ and ubiquitous computing is an exciting _________________ that promises to provide computing and communication services to the end users all the time and everywhere. Its systems are now invading in every aspect of our daily life and promise to revolutionize our life much more profoundly than elevators, electric motors or even personal computer evolution ever did. The emergence of this technology is a natural outcome of research and _________________ advances in a variety of areas including embedded systems, pervasive computing and communications, wireless networks, mobile computing, distributed computing and agent technologies.

paradigm, embedded, technological

6) _________________, or calm technology, is a paradigm shift where technology becomes virtually _________________ in our lives. Instead of having a desk-top or lap-top machine, the technology we use will be embedded in our _________________.

ubiquitous computing, environment, invisible

7) The idea behind ________________ is to surround ourselves with computers and software that are carefully tuned to offer us unobtrusive ________________ as we navigate through our work and ________________ lives.

assistance, ubiquitous computing, invisible

8) The any place/any time/any means vision of ________________has explosive impact on academics, industry, government and daily life. This emergence is the natural result of research and ________________ advances in wireless and sensor networks, embedded systems, mobile computing, distributed computing, agent technologies, autonomic computing and ________________.

ubiquitous computing, technological, communication)

9) ________________computing is the trend towards increasingly ubiquitous (another name for the movement is ubiquitous ________________, connected computing devices in the environment, a trend being brought about by a convergence of advanced electronic – and particularly, wireless – ________________ and the Internet.

technologies, pervasive, computing

10) At their core, all models of ________________ computing (also called pervasive computing) share a vision of small, inexpensive, robust networked processing ________________, distributed at all scales throughout ________________ life and generally turned to distinctly common-place ends.

devices, everyday, ubiquitous

2

Convergence

2 Convergence

2.1 Introduction

Original diagram © 2004 Nathan Shedroff

In the absence of a more specific context, convergence **denotes** the approach toward a **definite** value, as time goes on; or to a definite point, a common view or opinion, or toward a fixed or **equilibrium** state. Convergent is the **adjectival** form, and also a noun meaning an **iterative approximation.**

이번 장에서 꼭 알아야 할 단어

denotes
의미하다, 나타내다

If one thing ***denotes*** *another, it is a sign or indication of it.*
- Red eyes **denote** tiredness.
- While warm colors **denote** violence, cool colors promote a soothing effect.

definite
확실한, 정확한

If something is ***definite,*** *it is firm and clear, and unlikely to be changed.*
- It is too soon to give a **definite** answer.
- I don't have **definite** plans, but I'm hoping to visit the museum while I'm there.

equilibrium
평형(상태), 마음의 평정

Equilibrium *is a state of balance or a state of mental calmness.*
- She fainted because she lost her sense of **equilibrium.**
- Yoga is said to restore one's inner **equilibrium.**

adjectival
형용사의,
형용사적 기능의

Adjectival *may refer to Adjective, a part of speech that modifies a noun or a pronoun Adjectival noun or sometimes "adjectival", a noun that functions as an adjective.*
☞ *adjetival은 형용사를 일컫는 표현이다. 명사나 대명사를 수식해주거나 명사라도 문장 내 형용사처럼 사용될 경우 adjectival이라고 쓸 수 있다.*

Iterative
되풀이되는, 반복의

An ***iterative*** *p;rocess is a process where the same procedure is repeated over and over again to obtain a desired result = doing something again and again, usually to improve it.*
- **Iterative** processing of information during sleep may improve consolidation.
- This process can often be **iterative** and very time-consuming.

approximation
근사치

An ***approximation*** *is a guess of a number that is not exact but that is close.*
- I can't tell you exactly, but six meters is a close **approximation.**
- Could you give me a rough **approximation** of how many people will be coming?

Common Collocations

definite	equilibrium
definite plan definite answer definite about~	chemical equilibrium economic equilibrium sense of equilibrium
iterative	**approximation**
iterative design iterative process iterative development	a close approximation a rough approximation of ~

2.2 Useful Expressions

in the absence of : ~이 없을 때에, ~이 없으니까	
• In the absence of concrete evidence, we had to drop the charges. • In the absence of more suitable candidates, we decided to offer the job to Mr. Conway. • In the absence of a manager, staff must make their own decisions.	• 구체적인 증거가 없어서 우리는 기소를 취하해야 했다. • 더 적합한 후보가 없었기 때문에, 우리는 Conway씨에게 일할 기회를 주기로 결정했다. • 매니저가 없었기 때문에, 직원들은 스스로 결정을 내려야 했다.
approach toward : ~에 대한 접근	
• The pilot began her approach toward the runway. • The president is shifting his approach toward North Korea. • We will present about a new approach toward food.	• 그 조종사는 활주로를 향해 진입을 시작했다. • 대통령은 북한에 대한 접근방식을 바꾸고 있다. • 우리는 음식에 대한 새로운 접근법을 제시할 것이다.

2.3 Descriptive Grammar

… as time goes on ; or to a definite point…

Semicolon (;)은 주로 설명을 덧붙이는 경우에 쓰이며, 이때 양쪽의 문법적 형태는 같다. 또한 세미콜론은 접속부사와 함께 잘 쓰인다. 예시를 통해 세미콜론의 쓰임을 더 살펴보도록 하자.
Some people write with a word processor; others write with a pen or pencil.
(어떤 사람들은 워드 프로세서를 이용하여 글을 쓴다; 다른 이들은 펜이나 연필로 글을 쓴다.)

Most singers gain fame through hard work and dedication; however, Evita, found other means.
(대부분의 가수들은 근면과 헌신을 통해 명성을 얻는다; 그러나, Evita는 다른 방법을 찾았다.)

My grandmother seldom goes to bed this early; she's afraid she'll miss out on something.
(나의 할머니는 좀처럼 일찍 잠자리에 들지 않으신다; 그녀는 무언가를 놓치게 될까봐 두려워 하신다.)

위의 문장을 다시 살펴보면,

… as time goes on; or to a definite point …
시간이 지나면서 ; 어떤 명확한 지점에 다다르면서…

2.4 Substance

융합 또는 컨버전스(Convergence)라 함은 서로 다른 고객가치를 제공하던 기술 또는 제품들로부터 유사한 고객가치를 제공하고, 다른 기술 또는 제품들이 결합하여 새로운 고객가치를 창출되면서 기존 시장영역간의 경계가 불분명해지는 현상을 말한다.

영어 작문 연습

- Fusion or Convergence is the phenomena in which the techniques or products which have offered different costume values from each other supply similar costume value and can also create other new costume values combining with each until the former boundary among market areas become unclear.

Topics related to IT-based Convergence Technology and Service :

- Next Generation Communications and Networking
- Data/Information Processing Engineering on High Performance Computing
- Secure Information System and its Applications
- Intelligent Approaches on Information System
- Multimedia Technology and its Applications
- Broadcast Convergence Network and its Applications
- Embedded Systems and Software
- Telematics and Home Network Technology and its Applications
- ITS
- Game and Animation

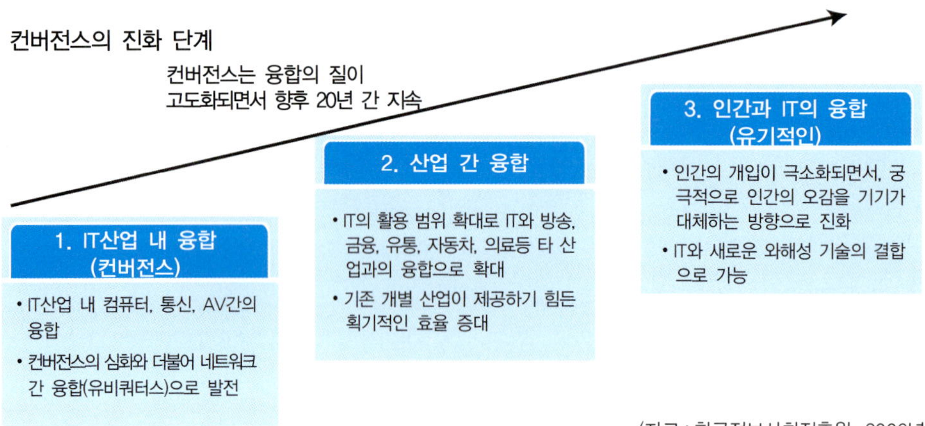

(자료 : 한국정보사회진흥원, 2006년)

[그림 4] 컨버전스의 진화 단계

이런 컨버전스를 통해 새로운 비즈니스 모델이 개발되고, 다양한 경로로 디지털 컨텐츠가 전달될 수 있다는 의미이다. 디지털 컨텐츠인 음성, 문자, 그림, 동영상 등의 멀티미디어 컨턴츠가 유선과 무선이 통합된 네트워크 유형으로 전달되기 때문에 유비쿼터스라는 하나의 패러다임의 기술적 동력이 되었다고 본다.

영어 작문 연습

- Ubiquitous Computing requires the following 4 conditions.
  It means that through such convergence a new business model is invented and digital contents can be transferred on various routs. For the Digital contents of voice, letters, pictures, and movies, multimedia are transferred with combined types of wired and wireless networks. Thus, convergence seems to become a kind of technical power of Ubiquitous computing.

2.5 Glossary

- Telecommunication convergence
- Enterprise FMC
- James John Bell
- Next Generation Networks
- Technological singularity
- Super intelligence
- Voice over IP

References [참고문헌]

[1] 유비쿼터스 개론 개념과 기술, 손병희 저, ITC 출판사, 2009
[2] 융합 그리고 넥스트, 손병희 저, 교우사, 2015
[3] Wikipedia
[4] Cambridge Advanced Learner's Dictionary, Cambridge University Press
[5] Collins COBUILD Intermediate Learner's Dictionary, Heinle & Heinle Publishers
[6] Longman Essential Activator, Longman National Corpus
[7] English Grammar in Use, Raymond Murphy, Cambridge University Press

2과 연습문제

01 Match each word in the first column with its definition in the second column.

1.	denote	A.	represent something
2.	definite	B.	a state of balance
3.	equilibrium	C.	Firm and clear
4.	iterative	D.	a guess of a number that is close
5.	approximation	E.	repeated over and over again
6.	in the absence of	F.	without
7.	approach toward	G.	coming near

01 Each of the following excerpts is about ubiquitous computing. Read the sentences. Fill in the blanks using the words in the below box.

1) Computational video includes the study and ________________ of the processing of streamed video data. This field of research is emerging from the ________________of two technologies: digital cameras and high performance computing and high bandwidth ________________.

convergence, networks, application

2) Most people appear to think that the future of Digital Media is ________________, that it is ________________that TVs and Computers are going to ________________and become a single device. Some people however think that this will not happen, and they may have good reason for thinking so.

convergence, converge, inevitable

3) ________________ is the coming together of two or more disparate disciplines or ________________. For example, the so-called fax revolution was produced by a convergence of ________________ technology, optical scanning technology, and printing technology.

telecommunications, convergence, technologies

3) In graphics, ________________ refers to how sharply an individual color pixel on a monitor appears. Each pixel is ________________ of three dots -- a red, blue, and green one. If the dots are badly misconverged, the pixel will appear blurry. All monitors have some convergence errors, but they ________________ in degree.

convergence, differ, composed

4) A computer convergence system includes a convergence functionality module, a computer, and a display ________________. The ________________ functionality module includes a first input for receiving a first video signal and a second input for ________________ a second video signal.

receiving, device, convergence

5) With the ________________ of the Web as a publishing and programming medium, as well as the availability of high-level development tools, the distinctions between algorithm and presentation are blurring. Today, a programmer needs to understand and apply ________________ of graphic design in order to develop applications that are usable and attractive to the user. Likewise, a graphic designer developing electronic media must understand and apply programming principles to control the dynamic behavior of the media. Today I will address the ________________ of programming and graphic design, from both the perspectives of a computer scientist and a graphic designer.

emergence, convergence, principles

6) Digital fusion – multimedia* – desktop media – is described as the ______________ of computing, television, audio, printing and telecommunications. Each of the ______________ has been significant. Bringing them together results in the whole having greater ______________ than each individual part and is one of the industry's most significant developments.

convergence, components, impact

7) The ______________ of digital ______________ and their use will impact the future of ______________, distance education and business.

teleconferencing, convergence, technologies

8) ______________ can be defined as the integration of telephony, computing and media technologies, and hence the ______________ of the businesses, markets and the social ______________ associated with them.

integration, interaction, convergence

9) Technological ______________ is already a common phenomenon in today's ICT devices. Computers are able to function as video and audio players, even having editing capabilities which is traditionally the area of professional broadcasting. The ______________ of internet connectivity, mobile telephony, computing power, media storage, digital camera, mobile gaming and digital audio player are already apparent in many of today's devices such as ______________ phones and personal digital assistants.

convergence, mobile, integration

3

Home Networks

3 Home Networks

3.1 Introduction

A home network is a **residential** local area network, and is used to connect multiple devices within the home.

The simplest home networks are used to connect 2 or more PCs for sharing files, printers, and a single connection to the Internet (usually **broadband** Internet through a cable or DSL provider). A server can be added for increased functionality.

More recently telephone companies such as AT&T and British Telecom have been using home networking to provide **triple** play services (voice, video and data) to customers. These use IPTV to provide the video service. The home network usually operates over the existing home wiring (**coax** in North America, phone wires in multi dwelling units (MDU) and powerline in Europe). These home networks are often professionally **installed** and managed by the telco. The ITU-T G.hn standard, which provides high-speed (up to 1 Gbit/s) local area networking over existing home wiring (power lines, phone lines and coaxial cables), is an example of a home networking technology **designed** specifically **for** IPTV delivery.

- From Wikipedia, the free encyclopedia

이번 장에서 꼭 알아야 할 단어

residential

주택지의

*A **residential** area contains houses rather than offices or factories.*

- Officetels are often not located in **residential** neighborhoods.
- Our company provides the cheapest and all around best **residential** computer repair services in Chungnam province.

broadband

광대역

__Broadband__ is a method of sending many electronic messages at the same time by using a wide range of frequencies.

- How many different companies offer **broadband** in this area?
  - ☞ *__Broadband in telecommunications__ refers to a signaling method that includes or handles a relatively wide range (or band) of frequencies, which may be divided into channels or frequency bins.*
  - ☞ *__Broadband in data__ can refer to broadband networks or broadband Internet and may have the same meaning as above, so that data transmission over a fiber optic cable would be referred to as broadband as compared to a telephone modem operating at 56,000 bits per second.*

triple

3 배의

__Triple__ means having three things or parts, or happening three times

- Her trainer is a **triple** Olympic chapion.
- There's a **triple** bill of Scorcese films (= three films) on at the cinema next Sunday.
  - ☞ ***Triple Play Service***

    TPS (Triple Play Service)가 제공이 되는데, 이는 한 번에 주자 3명을 아웃 시키는 야구 용어에서 착안해 만들어진 말로 방송과 초고속인터넷, 유선전화의 세가지 서비스를 한꺼번에 묶어 한 사업자가 모두 제공하는 것을 의미한다. 요즘은 QPS (Quadruple Play Service)라고 TPS + 이동통신서비스 네 가지를 한꺼번에 묶어 한 사업자가 모두 제공하는 서비스가 나오고 있는 시점이다.

- In telecommunications, the ***triple play service*** is a marketing term for the provisioning of two bandwidth-intensive services, high-speed Internet access and television, and a less bandwidth- demanding (but more latency-sensitive) service, telephone, over a single broadband connection.
- ***triple-play network*** is one in which voice, video and data are all provided in a single access subscription.

  Adding new triple play services to the service bundle has acompelled service providers to rethink their network infrastructure to factor in stringent requirements in terms of high availability, service flexibility and richness, scale, and service reach, as well as network and service manageability

coax

(고주파신호) 전송라인

Coax is a transmission line for high-frequency signals

- **coax cable**

동사로 사용되는 경우 의미에 주의하자!

*If you **coax** someone **to** do something, you persuade someone gently to do something, or go somewhere.*

- Perhaps you could **coax** your father into taking you to the station.
- A mother was **coaxing** her reluctant child **into** the water.
- He has some information I want, so I'm going to try to **coax** it out of him over a drink.

install

~을 설치하다

*If you **install** a piece of equipment, you fit it or put it somewhere so that it is ready to be used.*

- They **install** a new phone line.
- Can you help me **install** this software?

be designed for

~을 위해 계획되다 (만들어 지다)

*If something **is designed for** a certain purpose, it is created or done for that purpose.*

- This space was clearly not **designed for** that purpose.
- Shop for accessories, bags, storage and other products **designed for** our Samsung computers.
- FontFun is an interactive application **designed for** searching and exploring font databases.

Common Collocations

Residential	Broadband
Residential listings Residential school Residential mortgage Residential gateway Residential lease Residential elevator	Broadband speed test Broadband service Mobile broadband Wireless broadband Broadband internet Broadband availability
Triple Play	**ambient**
Triple play telecommunications Triple play broadband Triple play telecom Triple play providers Triple play IPTV	Coax connectors Coax splitter Digital coax Wireless coax Coax switch Coax amplifier

3.2 Useful Expressions

be added for : ~을 위해 더해지다, 추가되다	
• The whole wheat flour will be added for nutritional value. • Additional activities will be added for your enjoyment. • More choices can be added for your selection if this is not enough.	• 통밀가루는 영양가를 위해 더해질 것이다. • 추가적인 활동이 너의 즐거움을 위해 더해질 것이다. • 만약 이것이 충분하지 않다면 더 많은 선택안들이 추가될 것이다.
be managed by : ~에 의해 운영(관리)되다	
• The new project will eventually be managed by Koreans. • These employees are so lazy that they couldn't be managed by anyone! • It is difficult when foreigners are managed by Koreans because of cultural differences.	• 새로운 프로젝트는 결국 한국인들에 의해 운영될 것이다. • 직원들이 너무 게으르기 때문에 누구도 그들을 관리할 수 없었다. • 한국인이 외국인을 관리하는 것은 문화적 차이가 있기 때문에 어렵다.

3.3 Descriptive Grammar

More recently telephone companies *have been using* home networking.

현재완료진행형은 과거의 어느 시점에서 시작된 동작이 현재까지 계속된 경우에 쓰이며, 형태는 have(has) + been + -ing를 사용한다. 현재완료와 현재완료 진행 모두 과거부터 현재까지 동작이 계속되는 것을 나타내지만, 현재완료진행형은 동작이 지금까지 계속되고 있음을 강조한다. 아래 두 문장을 비교해 보자.

I *have written* a novel since last week. (현재완료)

I *have been writing* a novel since last week. (현재완료진행)

☞ (1)번 문장은 과거의 한 시점(지난 주)에 시작한 소설쓰기가 계속 이어지다가 이제(현재) 완료되었다는 의미라면, (2)번 문장은 지난 주부터 시작된 소설쓰기가 지금도 계속 된다는 의미이다. 예문을 더 살펴보도록 하자.

They *have been attending* university together for three years.

I *have been talking* to you for over an hour.

We'*ve been seeing* each other for 2 years.

She has been working on this project for two weeks.

What have you been doing for the past hour?

위의 문장을 다시 살펴보면,

More recently / telephone companies / *have been using* home networking.
최근 / 통신기업들이 / (예전부터 지금까지) 계속 사용하고 있다 / 홈네트워킹을

3.4 Substance

왜 홈 네트워크라고 할까? 라는 의문을 제기해 본다.
결론적으로 얘기하면, 네트워크의 이동이다.

영어 작문 연습

- We can ask the question "why home network?'
  To conclude, it is a movement of network.

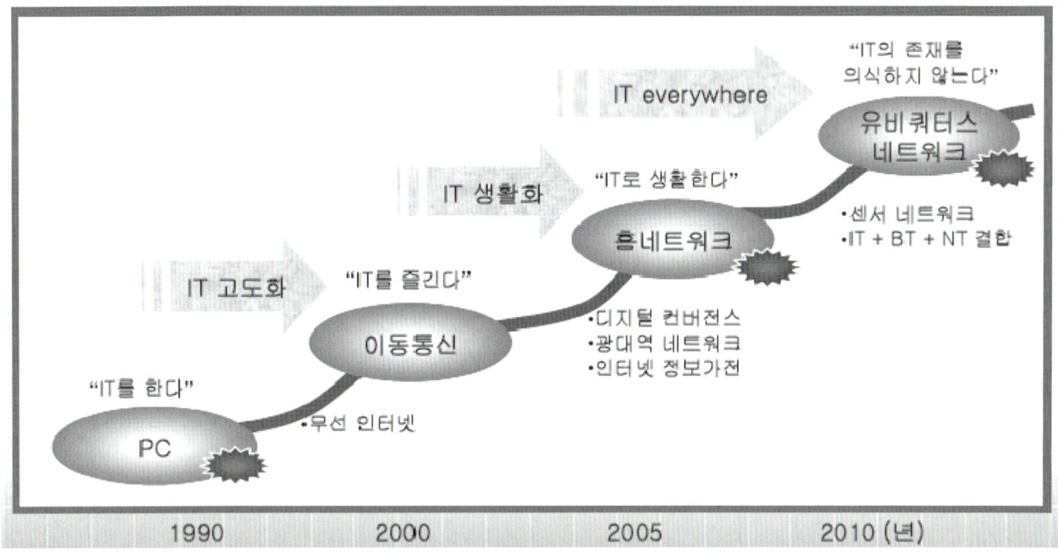

[그림 5] **네트워크의 이동**

이 대답은 1990년 PC의 보급에서부터 출발한다. 유비쿼터스의 탄생 배경과 맞물리는 시간이다. PC의 보급이 증가하면서 우리는 "IT를 한다."라고 얘기 했다. PC인 개인용 단말기는 그 형태가 다양해져 PDA, 노트북, 휴대폰, 내비게이션으로 이어 지는데, 나열한 단말기들의 공통점은 이동성을 보장한다는 점이다.

유선으로 연결되었던 네트워크가 자유를 얻은 셈이다. 무선 통신의 시대가 열리면서 우리는 "IT를 즐긴다."라고 얘기를 했다. 이때, 우리나라에서도 한 TV 스타가 어느 바 (Bar)에서 헤드 셋을 끼고, 음악을 들으면서 춤을 추는데 전화가 울리자, 머리를 옆으로 숙이며 귀에 있는 헤드 셋을 어깨로 툭 치며 통화를 하는 광고가 나오기도 했었다.

유무선 네트워크를 사용해서 IT를 즐기는 시간을 지나 다음에 나타난 현상은 유무선 네트워크들이 집으로 이동하게 된다. 비로소 "IT가 생활화 된다."라고 할 수 있게 된다.

가정 내 가전기기 및 시스템을 상호 또는 외부의 정보 기기와 연결하여 원격접근 및 제어가 가능하고, 음악, 비디오, 데이터 등과 같은 컨텐츠를 사용할 수 있도록 양방향 통신 서비스 환경을 구현하는 기술로 유무선 네트워크를 통하여 누구나 기기, 시간, 장소에 구애 받지 않고 다양한 서비스를 제공받을 수 있는 기술을 포함하고 있다. 여기서 우리는 홈을 편리한 홈, 안전하고 건강한 홈, 효율적인 홈, 즐거운 홈으로 종류를 분류할 수 있다.

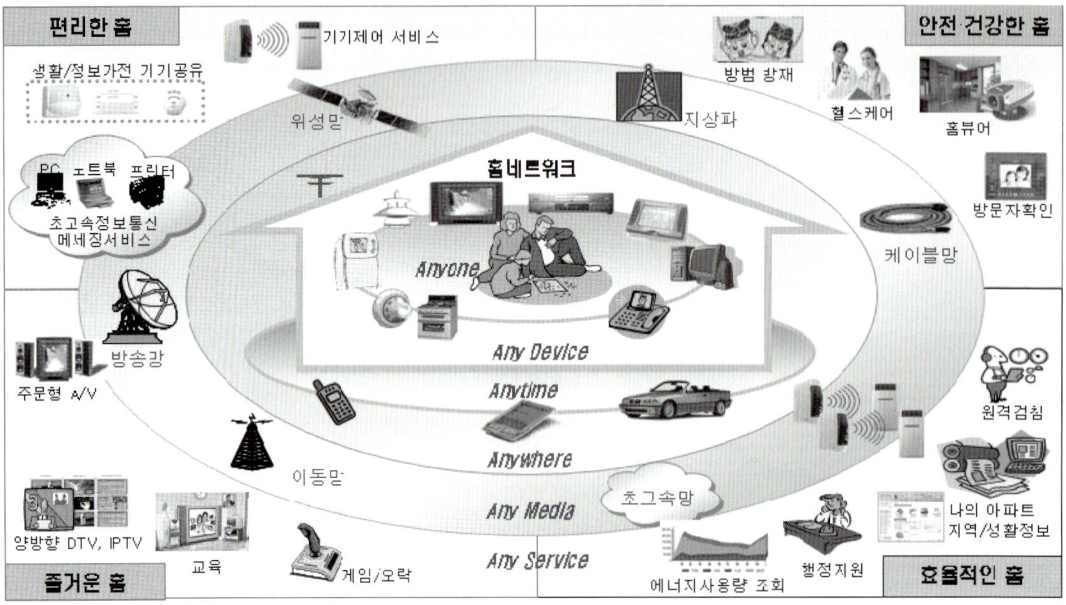

[그림 6] 홈 네트워크의 개념도

- 편리한 홈 : 사용자는 시간과 장소에 구애 받지 않고 정보가전기기를 이용하여 다양한 서비스를 제공받을 수 있는 미래지향적인 가정환경이다.
- 안전·건강한 홈 : 남녀노소 누구나 특별한 교육 없이도 정보·가전기기를 통하여 원격교육 및 원격의료 등의 복지서비스 혜택이 가능한 가정환경이다.
- 효율적인 홈 : 통신 인프라 구축의 성과를 극대화한 서비스 활성화 및 국민의 IT 생활화가 구현된 가정환경이다.
- 즐거운 홈 : 가정 내 네트워크와 외부의 유무선 네트워크를 통하여 디지털TV, 지능형로봇 및 디지털 컨텐츠 등 타 산업 분야와 융합된 다양한 서비스를 쉽게 이용할 수 있는 가정환경이다.

영어 작문 연습

- Convenient Home : home environment for the future where the user can be supplied with various services by using information appliances without limit of time and place.
- Safe and Healthful Home : home environment where people of all ages and both sexes can receive welfare service such as remote education and remote medical service without special instruction
- Effective Home : home environment where the service can be vitalized with maximized telecommunication infra construction result and IT in ordinary life can be embodied
- Pleasant Home : home environment where digital TV, intelligent robot, and digital contents and etc. fused with other industrial field can be used easily through network in home and wired or wireless network.

3.5 Glossary

A home network may consist of the following components:

- A broadband modem for connection to the internet (either a DSL modem using the phone line, or cable modem using the cable internet connection).
- A residential gateway (sometimes called a router) connected between the broad band modem and the rest of the network. This enables multiple devices to connect to the internet simultaneously. Residential gateways, hubs/switches, DSL modems, and wireless access points are often combined.
- A PC, or multiple PCs including laptops
- A wireless access point, usually implemented as a feature rather than a separate box, for connecting wireless devices
- Entertainment peripherals – an increasing number of devices can be connected to the home network, including DVRs like TiVo, digital audio players, games machines, stereo system, and IP set-top box.
- Internet Phones (VoIP)
- A network bridge connects two networks together, often giving a wired device, e.g. Xbox, access to a wireless network.
- A network hub/switch – a central networking hub containing a number of Ethernet ports for connecting multiple networked devices
- A network attached storage (NAS) device can be used for storage on the network.
- A print server can be used to share printers among computers on the network.

Older devices may not have the appropriate connector to the network. USB and PCI network controllers can be installed in some devices to allow them to connect to networks.

Network devices may also be configured from a computer. For example, broadband modems are often configured through a web client on a networked PC. As networking technology evolves, more electronic devices and home appliances are becoming Internet ready and accessible through the home network. Set-top boxes from cable TV providers already have USB and Ethernet ports "for future use".

References [참고문헌]

[1] 융합 그리고 넥스트, 손병희 저, 교우사, 2015
[2] Cambridge Advanced Learner's Dictionary, Cambridge University Press
[3] Collins COBUILD Intermediate Learner's Dictionary, Heinle & Heinle Publishers
[4] Longman Essential Activator, Longman National Corpus
[5] English Grammar in Use, Raymond Murphy, Cambridge University Press

3과 연습문제

01 Match each word in the first column with its definition in the second column.

1. triple	A. fit or put something somewhere
2. install	B. area filled with houses
3. residential	C. three things or parts, or happening three times
4. be designed for	D. be made for

02 Each of the following excerpts is about home networks. Read the sentences. Fill in the blanks using the words in the below box.

1) Once, ____________________ were primarily the realm of technophiles -- most families either didn't need or couldn't afford more than one computer. But now, in addition to using computers for e-mail, people use them for ____________________, shopping, instant messaging, downloading music and videos, and playing games. For many families, one ____________________is no longer enough to go around. In a household with multiple computers, a home network often becomes a necessity rather than a technical toy.

schoolwork, home networks, computer

2) A home network or home area network (HAN) is a residential local area ____________________. It is used for communication between digital devices typically deployed in the home, usually a small number of personal computers and ____________________, such as printers and mobile computing devices. An important function is the sharing of Internet access, often a broadband service through a cable TV or

Digital Subscriber Line (DSL) ____________________.

accessories, provider, network

3) ____________________can be set up with or without Internet ____________________, using various ____________________of computer hardware and software.

access, home networks, combinations

4) Most home ____________________are cobbled together without a lot of forethought about which devices are used to create it. For most, the decision is basically to use whatever their service ____________________shipped in a cardboard box. Whether they're going for wireless or ____________________, homeowners often haven't thought through their options.

networks, wired, provider

5) Increasingly, consumers are ____________________their stereos, plasma TVs, DVRs, PCs, laptops and gaming systems to ____________________entertainment systems designed to record, store and share media across devices throughout the home. As the home ____________________market continues to gain traction, what devices will they hook up next?

networking, connecting, intelligent

6) A computer network may fail to ____________________ properly for many different reasons. Understanding basic network ____________________methods in various situations will help you recover quickly on ____________________and elsewhere.

home networks, function, troubleshooting

7) Have you had trouble with ____________________ two computers? It's possible to set up a simple peer-to-peer network with two ____________________ in just a few minutes. Unfortunately, various ____________________ issues may arise that make the task much more difficult.

networking, computers, technical

8) If you have a ____________________, you'll welcome the easy file ____________________, remote access and the image-based backup ____________________ of Windows Home Server.

sharing, home network, features

9) Once you've designed and ____________________ your home network to suit your ____________________, it's tempting to think your work is done. But you still have to make sure your systems are ____________________, up to date, and running as trouble free as possible. You also need to know what to do when something goes wrong—which it almost certainly will.

configured, optimized, needs

10) Home ____________________ are increasingly becoming a part of everyday life, thanks to their convenience and the growth of supporting ____________________ such as DSL and cable modem connections. Learn how to determine the best type of Internet connection for your needs, how to select a reliable provider, and how to set up your ____________________ for maximum efficiency—whether wired or wireless.

technologies, network, offices

4

RFID

4 RFID

4.1 Introduction

Radio-frequency **identification** (RFID) is the use of an object (typically **referred to as** an RFID tag) applied to or incorporated into a product, animal, or person for the purpose of identification and tracking using radio **waves**. Some tags can be read from several meters away and beyond the line of sight of the reader.

Most RFID tags contain at least two parts. One is an integrated **circuit** for storing and processing information, **modulating** and demodulating a radio-frequency (RF) signal, and other specialized functions. The second is an antenna for receiving and **transmitting** the signal.

There are generally two types of RFID tags: active RFID tags, which contain a battery and can transmit signals **autonomously,** and passive RFID tags, which have no battery and require an external source to **provoke** signal transmission.

Today, RFID is used in enterprise supply chain management to improve the efficiency of inventory tracking and management.

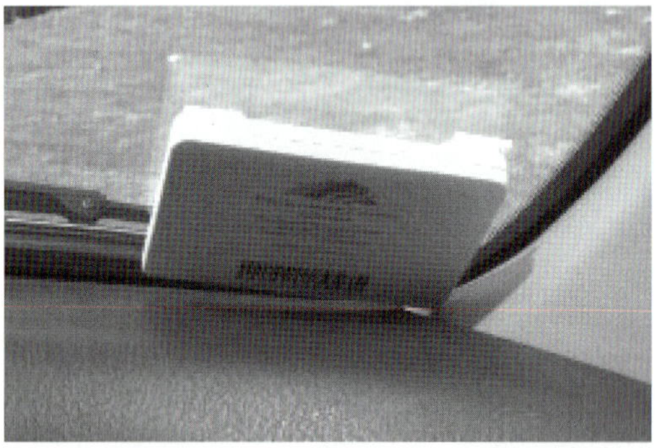

- From Wikipedia, the free encyclopedia

〈표 1〉 RF 주파수 대역별 분류

전파종류	주파수 대역	전파방식	적용분야
VLF	3KHz~30KHz	지표면 전파	장거리 무선항해, 해저통신
LF	30KHz~300KHz		장거리 무선항해, 항해위치 확인기
MF	300KHz~3MHz	대류권 전파	AM라디오, 긴급구조주파수
HF	3MHz~30MHz	전리층 전파	아마추어라디오, 군사통신, 장거리 항공기
VHF	30MHz~300MHz	가시거리 전파	VHF 텔레비전, 이동전화, 셀 방식라디오
UHF	300MHz~3GHz		UHF 텔레비전, 이동전화, 위성 마이크로파, 레이더 통신
SHF	3GHz~30GHz	우주공간 전파	지상 마이크로파, 위성 마이크로파, 레이더 통신
EHF	30GHz~300GHz		레이더, 위성 통신

유비쿼터스 기술을 이해하는 데 도움이 되는 팁

무선 주파수 분류

무선 주파수 대역이 처음에는 복잡해 보일지 모르나, 종류의 풀 네임을 보면 쉽게 이해할 수 있다. 우선 LF(Low Frequency), MF(Medium Frequency), HF(High Frequency) 3가지로 저주파, 중주파, 고주파로 분류된다. 이를 다시 세분화 하여 Very, Ultra, Super, Extremely로 표현한 것이다. 주로 실생활에서 이용되는 주파수 대역은 VHF와 UHF를 사용한다.

이번 장에서 꼭 알아야 할 단어

identification

확인, 식별

Identification is an official document that shows or proves who you are.

- To access your funds at the bank, you must enter your personal **identification** number twice.
- Most of the bodies were badly burned, making **identification** almost impossible.
- We were asked to show some **identification** before the security guards would let us in.

refer to

~라고 부르다

If you refer to someone or something, you mention them.

- Even though I'm Korean, people still **refer to** me as Chinese.
- I'm not really the boss, but people **refer to** me as one.

☞ *refer to A as B라고 하면 'A를 B라고 부르다' 라는 뜻이지만 대개 수동형으로 사용되어 A is referred to as B라는 형태로 많이 사용된다.*

wave

파동, 파장

A wave is the pattern in which some types of energy, such as sound, light, and heat, are spread or carried.

- Radio **waves** have the longest wavelengths in the electromagnetic spectrum.
- Radio **waves** are that part of the electromagnetic spectrum with wavelengths longer than infrared light.
- Electromagnetic radiation is a ubiquitous phenomenon that takes the form of self-propagating **waves** in a vacuum or in matter.

circuit

회로

Circuit is a closed system of wires or pipes through which electricity or liquid can flow.

- A defect was found in the water-cooling/electrical **circuit.**
- An integrated **circuit** is a very small electronic circuit which consists of a lot of small parts made on a piece of semiconducting material.
- If those two wires touch, the appliance will short-**circuit** and probably go up in flames.

modulate

(전파, 신로를) 맞추다, 조절하다

*To **modulate** means to mix an electrical signal which represents sounds or pictures with a radio signal so that it can be broadcast*

- A **modulating** furnace makes slight operating adjustments while achieving maximum efficiency.
- A **modulating** valve includes an inlet passage and a discharge passage, both of which are in communication with an internal cavity associated with the valve body.

transmit

~을 전송하다

*When a message or electronic signal is **transmitted,** it is sent by wires, radio waves, or satellite.*

- This is currently the most efficient way to **transmit** electronic mail.
- The information is **transmitted** electronically to the central computer.
- In what way is this disease **transmitted**?

autonomous

자유로운, 독립적인

: independent and having the power to make your own decisions
*adverb(부사) : **autonomously***

- The result is a lot of fun to drive around, and even more fun to watch it **autonomously** detect, target, and destroy loose LEGO targets.
- The central idea is that when someone chooses and acts **autonomously**, they are in control of their own choices and actions.
- Remotely operated underwater vehicles and **autonomously** operated underwater vehicles have found many applications, including inspection and repair of offshore oil rigs and ships, and the exploration of shipwrecks and underwater caves.

provoke

~을 자극하다

*If something **provokes** a certain reaction, it causes it.*

- The article in the newspaper **provoked** anger.
- Our report **provoked** some strong views and interests, and I welcome that.
- Does an interface with less assistance **provoke** more thoughtful behavior?

enterprise

기업, 사업

*An **enterprise** is a company or business*

- Don't forget this is a commercial **enterprise** - we're here to make money.
- The **enterprise** is planning to extend its business abroad.

inventory

재고(물품) 목록

*An **inventory** is a written list of all the objects in a place.*

- Do you want me to print out the **inventory** list?
- How do you keep track of **inventory**?
- I think we need to focus on ways to shrink **inventory.**

Common Collocations

Identification	Wave	Circuit
Fake identification Identification card Identification psychology Personal identification number	Neutrogena wave Crime wave Wave files Ocean wave	Electrical circuit Circuit training Electronic circuit Short circuit
Modulating	**Autonomous**	**Inventory**
Modulating boiler Modulating valve Modulating thermostat	Autonomous robot Autonomous nervous system Autonomous vehicles	Inventory accounting Inventory software Inventory management Inventory system

〈표 2〉 RFID 대체기술과의 비교표

항목	RFID	자기 카드	바코드
인식 방법	비접촉	접촉	비접촉
거리	수십 m	접촉	수십 cm
인식 속도	1s 이내	수초	수초
투과력	투과 가능	불가	불가
데이터	64Kbtye	수백 byte	100byte
인식률	99% 이상	95% 정도	95% 정도
수명	반영구적	훼손 쉬움	훼손 쉬움

유비쿼터스 기술을 이해하는 데 도움이 되는 팁

RFID의 목적
RFID는 AIDC(Automatic Identification and Data Capture)을 목적으로 사용되는데 이를 자동 식별 및 데이터 수집 기술이라고 부른다.

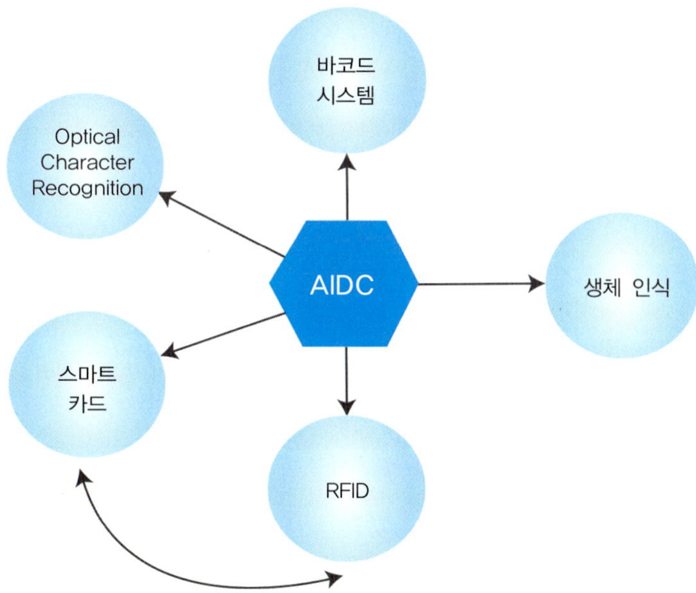

[그림 7] 자동 식별 및 데이터 수집 기술

AIDC는 자동으로 개체를 식별하고, 데이터를 수집하여, 시스템에 직접 데이터를 전하는 기술을 말한다. 일반적으로 AIDC 기술에는 바코드 시스템, RFID, 광학 문자 인식 시스템, 스마트 카드 등이 포함된다.
여기서 RFID와 스마트 카드는 대체 기술이기도 하다. 이 뿐만 아니라, AIDC 기술은 생체 인식 시스템에서도 사용이 되는데, 지문, 홍채, 또는 음성과 같은 특성을 식별하기도 한다.

4.2 Useful Expressions

the use of : ~의 사용(활용)	
• She lost the use of one arm. • She is physically challenged, so she paints with only the use of her mouth. • Your use of color in this painting is magnificent!	• 그녀는 한쪽 팔을 잃었다. • 그녀는 신체적으로 장애가 있다. 그래서 그녀는 오로지 입을 사용해서 그림을 그린다. • 이 그림에서 네가 사용한 색감이 대단한걸!
improve (or increase) the efficiency of~ : ~의 효율성을 높이다	
• A team system was introduced to improve the efficiency of the organization. • Let's brainstorm some ways to increase the efficiency of the system. • Due to recent improvements to the efficiency of workers, we are giving everyone a raise.	• 팀 제도가 조식의 효율성을 증진시키기 위해 도입되었다. • 이 시스템의 효율성을 증가시키기 위한 방안들을 생각해 보자. • 최근 직원들의 효율성의 향상으로 인해, 우리는 모두에게 월급을 인상해 주었다.

4.3 Descriptive Grammar

등위접속사는 and, but, or, yet 등이 있으며 접속사를 기준으로 단어와 단어, 구와 구, 절과 절을 대등하게 연결해 준다는 문법적 특성이 있다. 아래 예문을 통해 등위접속사의 특징을 확인해 보도록 하자.

- There were several cars *and* trucks in the street.
- I shopped at the store, *and* I went home.
- It rained last night, *and* the road was wet.
- You can have a soda *or* lemonade.
- We could walk, *or* take a taxi.
- We tried our best, *but* we lost the game.
- I'm tired, *yet* I can't sleep.

☞ 이번 장의 본문 내용이다. 〈보기〉 문장과 같이 접속사를 찾아 체크하고, 접속사를 중심으로 대등하게 연결된 단어나 어구를 찾아 밑줄을 그어보자.

보기

Some tags can be read from several meters away *and* beyond the line of sight of the reader.

- One is an integrated circuit for storing and processing information, modulating and demodulating a radio-frequency (RF) signal, and other specialized functions. The second is an antenna for receiving and transmitting the signal.
- There are generally two types of RFID tags: active RFID tags, which contain a battery and can transmit signals autonomously, and passive RFID tags, which have no battery and require an external source to provoke signal transmission.
- Today, RFID is used in enterprise supply chain management to improve the efficiency of inventory tracking and management.

4.4 Substance

Human implants

Hand with the planned location of the RFID chip.

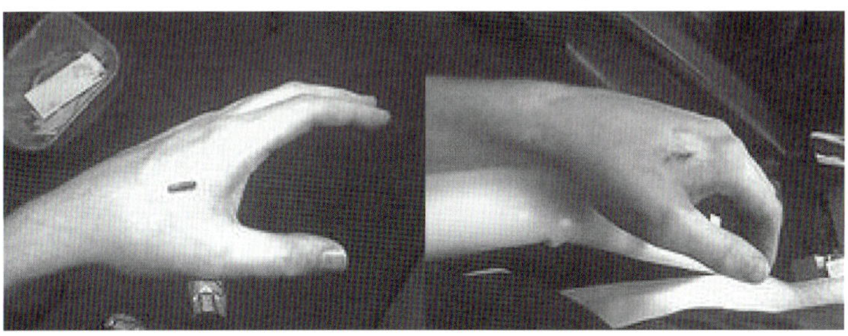

Just after the operation to insert the RFID tag was completed.

한글로 번역하기!!!!!

A human microchip implant is an integrated circuit device or RFID tag encased in silicate glass and implanted into a human's body. Such subdermal implants can be used for information storage, including personal identification, medical history, medications, allergies, and contact information.

영어 작문 연습

- A human microchip implant is an integrated circuit device or RFID tag encased in silicate glass and implanted into a human's body. Such subdermal implants can be used for information storage, including personal identification, medical history, medications, allergies, and contact information.

4.5 Glossary

- Biometrics
- Barcode
- Electronic Product Code
- E-ZPass
- Local Positioning Systems
- High Capacity Color Barcode
- List of emerging technologies
- Mass surveillance
- Microchip implant (human)
- Microchip implant (animal)
- Mobile RFID
- Near Field Communication
- Pharmacy informatics
- Polymer electrolyte
- Proximity card
- Identification friend or foe
- RFID race timing
- Real-time locating
- RuBee tags
- Smart cards
- Supranet
- Tracking system
- U-Key
- AS5678

References [참고문헌]

[1] 융합 그리고 넥스트, 손병희 저, 교우사, 2015
[2] http://www.conradchase.com
[3] Barcelona clubbers get chipped. BBC News, 29 Sep 2004
[4] http://www.spychips.com/press-releases/mexican-implant-correction.html
[5] High-tech cloning
[6] http://blogs.reuters.com/blog/2006/07/22/high-tech-cloning
[7] Vericip hacked press release, Spychips
[8] http://www.spychips.com/press-releases/verichip-hacked.html
[9] "Demo: Cloning a Verichip". http://cq.cx/verichip.pl. Retrieved on 2007-02-03.
[10] "VeriChips Implanted at City Watcher.com". Compliance and Privacy.
[11] http://www.complianceandprivacy.com/News-VeriChip-concern.asp. Retrieved on 2007-02-03. "No one I spoke with at Six Sigma Security or at CityWatcher knew that the Veri Chip had been hacked. They were also surprised to hear of Veri Chip's down sides as a medical device. It was clear they weren't aware of some of the controversy surrounding the implant. (Liz McIntyre)"
[12] Cambridge Advanced Learner's Dictionary, Cambridge University Press
[13] Collins COBUILD Intermediate Learner's Dictionary, Heinle & Heinle Publishers
[14] Longman Essential Activator, Longman National Corpus
[15] English Grammar in Use, Raymond Murphy, Cambridge University Press

4과 연습문제

01

Match each word in the first column with its definition in the second column.

1.	identification	A.	a closed system of wires or pipes
2.	refer to	B.	mention
3.	circuit	C.	send
4.	transmit	D.	an official document that shows or proves who you are
5.	autonomous	E.	independent
6.	provoke	F.	a written list
7.	enterprise	G.	a company or business
8.	inventory	H.	cause something

02

Each of the following excerpts is about RFID. Read the sentences. Fill in the blanks using the words in the box.

1) Radio-frequency identification ________________ is the use of an object (typically referred to as an RFID tag) applied to or incorporated into a product, animal, or person for the purpose of ________________ and tracking using radio waves. Some tags can be read from several ________________ away and beyond the line of sight of the reader.

meters, identification, RFID

2) Long checkout lines at the grocery store are one of the biggest complaints about the shopping experience. Soon, these lines could ________________ when the ubiquitous Universal Product Code (UPC) bar code is replaced by smart labels, also called radio frequency

identification (RFID) tags. ________ tags are intelligent bar codes that can talk to a ________ system to track every product that you put in your shopping cart.

RFID, labels, networked

3) Motorola's RFID solutions enable you to ________ your business to a new level of efficiency by providing greater visibility into your ________ as it moves across the supply chain. With real-time tracking information, you'll always know where your critical ________ assets are.

business, advance, inventory

4) The federal government now requires US citizens to present identity documents when entering the US. These documents incorporate ________ technology that jeopardizes the privacy and security of US ________. EPIC has previously urged the State Department to abandon the ________.

business, advance, inventory

5) Although there are lingering concerns by privacy advocates over the use of ________, there's no doubt that the use of RFID tags continues to increase. According to a new report from In-Stat, over 1 billion tags were produced last year, and by 2010, that total will rise to 33 billion. Wal-Mart's requiring of their suppliers to use RFID ________ are certainly one reason for the projected growth, as well as potential applications in both the industrial and retail sectors, such as the tracking of ________.

RFID, inventory, tags

6) ________________ is short for radio frequency identification, a technology similar in theory to bar code identification. With RFID, the electromagnetic or ________________ coupling in the RF portion of the electromagnetic spectrum is used to transmit signals. An RFID system consists of an antenna and a ________________, which read the radio frequency and transfer the information to a processing device, and a transponder, or tag, which is an integrated circuit containing the RF circuitry and information to be transmitted.

transceiver, electrostatic, RFID

7) Radio frequency identification (RFID) is a generic ________________ that is used to describe a system that ________________ the identity (in the form of a unique serial number) of an object or person wirelessly, using radio waves. It's grouped under the broad category of ________________ identification technologies.

automatic, term, transmits

8) RFID ________________, a technology once limited to tracking cattle, are tracking consumer products worldwide. Many manufacturers use the tags to track the ________________ of each product they make from the time it's made until it's pulled off the shelf and tossed in a shopping ________________.

location, tags, cart

9) RFID is in use all around us. If you have ever chipped your pet with an ID ________________, used EZPass through a toll booth, or paid for gas using SpeedPass, you've used ________________. In addition, RFID is increasingly used with biometric technologies for ________________.

tag, RFID, security

10) Today, RFID is used in enterprise supply chain ________________ to improve the efficiency of inventory tracking and management. However, a fair cost-sharing mechanism between retailer and consumer good suppliers is essential for a sustainable ________________ adoption to take place. In addition, it is also necessary for consumer good suppliers to be able to justify their investments in RFID ________________ with rational motives and justified returns.

RFID, management, technology

공대생을 위한 전공영어

5

WSN

5.1 Introduction
5.2 Useful Expressions
5.3 Descriptive Grammar
5.4 Substance
5.5 Glossary
연습문제

5 WSN

5.1 Introduction

A wireless sensor network (WSN) is a wireless network **consisting of** spatially distributed autonomous devices using sensors to cooperatively monitor physical or environmental conditions, such as temperature, sound, **vibration,** pressure, motion or pollutants, at different locations. [1][2] The development of wireless sensor networks was originally motivated by military **applications** such as battlefield **surveillance.** However, wireless sensor networks are now used in many industrial and civilian application areas, including industrial process monitoring and control, machine health monitoring, environment and habitat monitoring, healthcare applications, home automation, and traffic control.[1][3]

In addition to one or more sensors, each node in a sensor network is typically **equipped with** a radio transceiver or other wireless communications device, a small microcontroller, and an energy source, usually a battery. The envisaged size of a single sensor node can **vary from** shoebox-sized nodes down to devices the size of grain of dust, [1] although functioning 'motes' of genuine microscopic dimensions **have yet to** be created. The cost of sensor nodes is similarly variable, ranging from hundreds of pounds to a few pence, depending on the size of the sensor network and the complexity required of individual sensor nodes.[1] Size and cost **constraints on** sensor nodes result in corresponding **constraints** on resources such as energy, memory, computational speed and bandwidth.[1]

A sensor network normally constitutes a wireless ad-hoc network, meaning that each sensor supports a multi-hop routing algorithm (several nodes may forward data packets to the base station).

In computer science and telecommunications, wireless sensor networks are an active research area with numerous workshops and conferences arranged each year.

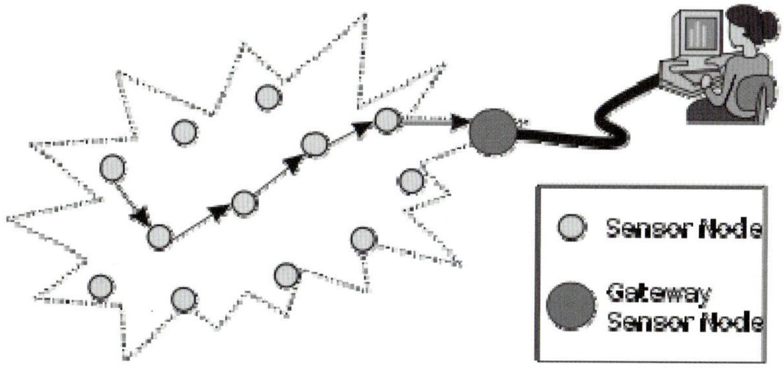

- From Wikipedia, the free encyclopedia

이번 장에서 꼭 알아야 할 단어

vibration
진동

*A **vibration** is a continuous, quick, slight shaking movement.*

- **Vibrations** were felt hundreds of miles from the centre of the earthquake.
- Aircraft manufacturers want to reduce **vibration** for the sake of safety.

surveillance
감시

***Surveillance** is the careful watching of a person or place, especially by the police or army, because of a crime that has happened or is expected.*

- The police have kept the nightclub under **surveillance** because of suspected illegal drug activity.
- More banks are now installing **surveillance** cameras.

application

응용 프로그램

*An **application** can refer to a computer program that is designed for a particular purpose.*

- The design has many **applications.**
- The **application** of this research has implications for the treatment of cancer.

equip with

장비를 갖추다

*If you **equip** a person or thing **with** something such as a tool or a machine, you provide them with it.*

- Their school is not well **equipped with** computers.
- Are you **equipped with** the skills necessary for the job?
- I am not **equipped with** the acumen to handle this problem.

vary

다양하다

*If things **vary,** they are different in size, amount, or degree.*

- Results may **vary.**
- You should **vary** your exercise so as not to put too much stress on your muscles.
- Prices may **vary** depending on the country you're in.

constraints on

~에 대한 제약

*A **constraint** is something that limits or controls what you can do*

- Financial **constraints** on the company are preventing them from employing new staff.
- By greatly reducing the hypothesis space, these **constraints** on hypotheses help render the inductive problem soluble.
- Can you tell me more about the legal **constraints** on the Executive Branch?

bandwidth

대역폭

*A **bandwidth** is the amount of information that can be sent between computers, through a telephone wire, etc.*

- The system will handle not only telephone calls and data messages but other signals that need high **bandwidth**, for instance those that encode TV pictures.
- high **bandwidth** services/applications

ad-hoc

즉석에서 마련된

*Something that is **ad hoc** is made or happening only for a particular purpose or need, not planned before it happens.*

- An **ad hoc** meeting will be held in five minutes.
- We deal with problems on **an ad** hoc basis.

ranging from A to B

A 에서 B 까지 범위에 걸쳐있는

The amount, number or type of something between an upper and a lower limit

- The price **range** is **from** $100 **to** $500.
- The age **range** is **from** eighteen **to** forty.
- Salaries **ranging from** $2,000 **to** $3,000 monthly are offered to entry level employees.

Common Collocations

Vibration	Surveillance	Application
Vibration radio Vibration meter Vibration therapy Vibration analysis	Surveillance system Surveillance camera Surveillance devices Surveillance equipment	Job application Application software iPhone applications Online application
constraints	**Bandwidth**	**ad–hoc**
Project constraints Oracle constraints Database constraints SQL constraints	Bandwidth test Bandwidth monitor Network bandwidth Computer bandwidth	Ad hoc network Ad hoc fallacy Ad hoc reports Ad hoc gallery

유비쿼터스 기술을 이해하는 데 도움이 되는 팁

RFID와 USN 기술의 개요
RFID (Radio Frequency Identification) / USN (Ubiquitous Sensor Network)은 모든 사물에 부착된 RFID 태그 또는 센서를 초소형 무선장치에 접목하여 이들 간의 네트워킹과 통신으로 실시간으로 정보를 획득, 처리, 활용하는 네트워크 시스템이다. 유비쿼터스의 핵심 기술이다.

5.2 Useful Expressions

Consist of : ~으로 구성되다, 이루어져 있다	
• The team consists of four Koreans and two Americans. • It's a simple dish to prepare, consisting mainly of rice and vegetables. • His diet consisted of bread, cheese, and water. • This cake consists mainly of sugar, flour, and butter.	• 그 팀은 네 명의 한국인과 두 명의 미국인으로 구성되어 있다. • 그것은 준비하기 간단한 요리이다, 주로 밥과 야채로 이루어진. • 그의 식단은 빵, 치즈, 물로 이루어져 있다. • 이 케익은 주로 설탕, 밀가루, 그리고 버터로 만들어졌다.
have yet to (do) : 아직 ~하지 않았다	
• Great things have yet to come. • I have written to your company several times, but I have yet to receive a reply. • We have yet to decide what action to take • The thing women have yet to learn is nobody gives you power; You just take it.	• 위대한 일들이 아직 일어나지 않았다. (앞으로 일어날 것이라는 뜻) • 저는 당신의 회사에게 여러 번 글을 썼지만, 아직 답변을 받지 못했습니다. • 우리는 어떤 조치를 취할지 아직 결정하지 않았다. • 여성들이 아직 알지 못하는 것은 그 누구도 당신에게 힘(권력)을 (아무 대가 없이) 주지 않는 다는 것이다.

5.3 Descriptive Grammar

The cost of sensor nodes is similarly variable, rangi*ng* from hundreds of pounds to a few pence, ...

분사란 동사에 ~ing나 ~ed를 붙여서 명사를 수식하는 형용사처럼 사용하는 것을 말한다. 즉, 동사에 ~ing나 ~ed의 옷을 입혀서 형용사로 사용하는 것이다. 이 때 ~ing를 붙이면 능동/진행의 의미가 그리고 ~ed의 경우에는 수동/완료의 어감이 더해진다. Give는 '~에게 -을 주다'라는 뜻을 가진 동사이다. 여기에 ~ing를 붙이면 'giving'이고, 능동의 의미가 더해져 '주는, 베푸는'이라는 뜻이 된다. 반면 '~ed'를 붙이면 given(give는 불규칙 동사이므로: give-gave-given) 주어가 무언가를 하는 것이 아닌 영향을 받는 수동의 의미가 더해져 '주어진, 받는'이라는 의미가 되는 것이다. 분사의 예를 하나 더 살펴보도록 하자.

Surprise (~를 놀라게 하다) - surprising (~를 놀라게 만드는) - surprised (놀란)

이러한 분사가 이끄는 구문을 분사구문이라고 한다. 대표적인 형태가 위 문장처럼 쉼표 (comma) 다음에 분사로 시작하는 구문이다.

- He was eating alone, *listening* to music.
- I forgot to call, *surprised* by the news.

위의 문장을 다시 살펴보면,

The cost of sensor nodes is similarly variable, ranging from hundreds of pounds to a few pence, …
센서노드의 가격이 다양하다 / 수백 파운드부터 몇 펜스까지 범위에 걸쳐 / …

[그림 8] 가정에서의 WSN

5.4 substance

RFID/USN은 어느 곳에나 부착된 태그와 센서 노드로부터 사물 및 환경 정보를 감지·저장·가공·통합하고 상황인식 정보 및 지식 컨텐츠 생성을 통하여 언제, 어디서, 누구나 원하는 맞춤형 지식 서비스를 자유로이 이용할 수 있는 첨단 지능형 네트워크이다.

영어 작문 연습

- RFID/USN is high-tech network enabling anyone to use knowledge service on demand anywhere and anytime freely by sensing. storing. processing. unifying object and environment information and producing contextual information and knowledge contents by tag and sense node attached anywhere.

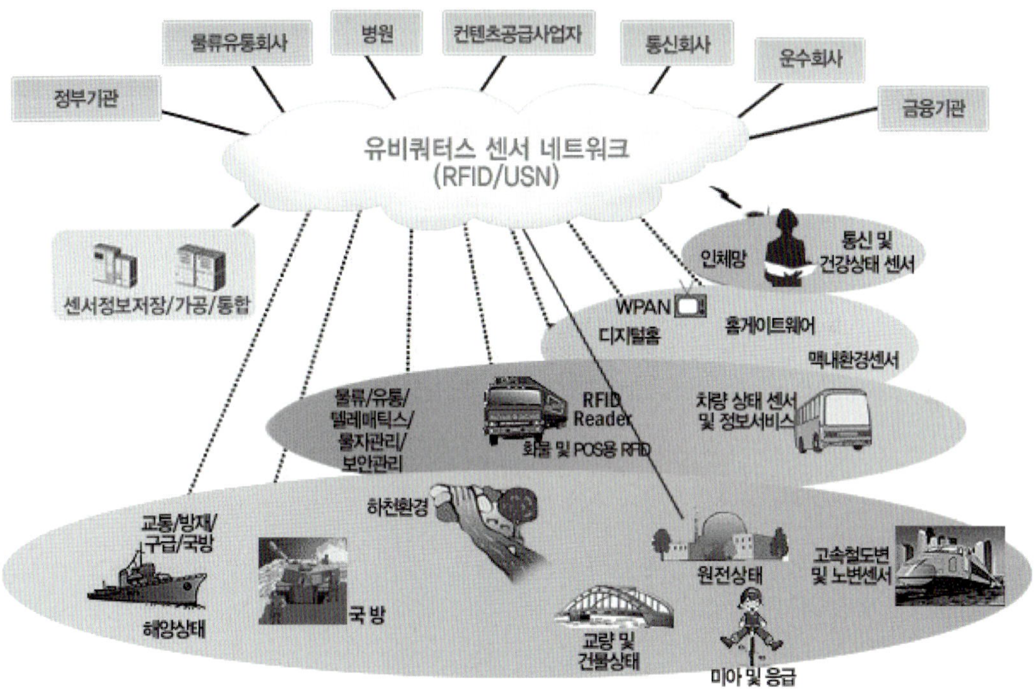

[그림 9] USN 개념

먼저 인식정보를 제공하는 RFID를 중심으로 발전하고 이에Sensing 기능이 추가되어 이들 간의 네트워크가 구축되는 USN 형태로 발전하고 있다.

영어 작문 연습

- First it is developing around RFID offering recognition information, and on this, with added sensing function it is developing in the form of USN constructing network among these.

5.5 Glossary

- Sensor node
- Mesh networking
- Mobile ad-hoc network (MANETS)
- neuRFon
- Smartdust
- Sensor Web
- TSMP
- Visual sensor network
- Key distribution in wireless sensor networks
- EnOcean
- Sun SPOT
- telemetry
- Location estimation in sensor networks
- Dust Networks

References [참고문헌]

[1] 융합 그리고 넥스트, 손병희 저, 교우사, 2015

[2] Römer, Kay; Friedemann Mattern (December 2004). "The Design Space of Wireless Sensor Networks". IEEE Wireless Communications 11 (6): 54-61. doi:10.1109/MWC.2004.1368897.
http://www.vs.inf.ethz.ch/publ/papers/wsn-designspace.pdf.

[3] Thomas Haenselmann (2006-04-05). Sensornetworks. GFDL Wireless Sensor Network textbook. http://www.informatik.uni-mannheim.de/~haensel/sn_book. Retrieved on 2006-08-29.

[4] Hadim, Salem; Nader Mohamed (2006). "Middleware Challenges and Approaches for Wireless Sensor Networks". IEEE Distributed Systems Online 7 (3): 1.doi:10.1109/MDSO.2006.19.
http://dsonline.computer.org/portal/pages/dsonline/2006/03/o3001.html. Art. No. 0603-o3001.

[5] Cambridge Advanced Learner's Dictionary, Cambridge University Press

[6] Collins COBUILD Intermediate Learner's Dictionary, Heinle & Heinle Publishers

[7] Longman Essential Activator, Longman National Corpus

[8] English Grammar in Use, Raymond Murphy, Cambridge University Press

5과 연습문제

01 Match each word in the first column with its definition in the second column.

1.	vibration	A.	the careful watching of a person or place
2.	surveillance	B.	provide with
3.	application	C.	quick, slight shaking movement
4.	equip with	D.	a computer program that is designed for a particular purpose
5.	vary	E.	different in size, amount, or degree
6.	constraints on	F.	not planned
7.	ad-hoc	G.	limits on

02 Each of the following excerpts is about WSNs. Read the sentences. Fill in the blanks using the words in the box.

1) A wireless sensor network (WSN) consists of spatially distributed autonomous ____________________ to cooperatively monitor physical or environmental conditions, such as temperature, sound, vibration, pressure, motion or pollutants. The development of ____________________ sensor networks was motivated by military applications such as battlefield surveillance. They are now used in many industrial and civilian ____________________ areas, including industrial process monitoring and control, machine health monitoring, environment and habitat monitoring, healthcare applications, home automation, and traffic control.

application, sensors, wireless

2) As ________________ sensors come onto the market and become a practical option, sensor networks take on a whole new dimension. The question arises: How can traditional ________________ topologies be adapted to the new ________________ medium?

network, communications, wireless

3) According to GainSpan, its technology provides the lowest power consuming Wi-Fi single chip ________________ for wireless sensor ________________ and other embedded applications, allowing devices to run for up to 10 years on a single AA ________ __ ________.

battery, solution, networks

4) ________________, otherwise known as Ubiquitous Sensor Networks (USN) will be used to automatically monitor building occupancy and climate control, oil and gas pumping, forest fires, avalanches, the effect and likely effect of hurricanes, traffic, hospitals and much more over wide areas over the coming years. The WSN ________________ will grow rapidly to exceed two billion dollars in yearly sales in the next ten years and about six billion dollars during the following ten years. In addition, new this year is coverage of energy harvesting and energy ________________ for small electronic devices.

WSN, storage, industry

5) A typical ________________ network device comprises the following components some of which are optional: power supply, microcontroller, ________________ communication, wired communication, sensor, local storage, and real time clock systems. The principal idea is that the sensors are connected to a tiny ________________ that coordinates the measurement, preprocesses,

stores and delivers the information.

computer, wireless, ensor

5) Researchers have long been promising that ________________ sensor networks will bring a new wave of smart appliances that can communicate with each other, and with central servers when they need attention. The ________________ for these appliances has existed for some time now. However, the commercial availability of such ________________ has been slow to market for one primary reason – the lack of structured business model.

products, technology, wireless

6) Who said Wireless Sensor Networks had to be small personal ________________ networks? According to Network World, researchers at Penn State University are developing a ________________ that would be attached to cars that would allow it to talk to other cars, forming a large area sensor ________________.

area, network, sensor

7) It was predicted back in 2005 that WSN would be a disruptive ________________ that changed the way we viewed the ________________. Today, market research still predicts that the ________________ revolution is "just around the corner".

world, technology, WSN

8) WSN is a great ________________, but that's just it. Its another technology. The technology alone doesn't do anything; it needs to be applied. And for most of the things that it can be applied to, there is often a cheaper, lower-________________, more reliable ________________.

method, technology, tech

9) As wireless ______________ become real commodities on the market, new options or new arguments for old options are causing professionals to consider ______________ strategies once ruled out. Let's look at the three classic network topologies (point-to-point, multidrop, and web), assess their strengths and weaknesses, and look at how the rules have changed now that ______________ systems are coming online.

wireless, network, sensors

6

Ad-hoc Networks

6 Ad-hoc Networks

6.1 Introduction

Since the **inception** of wireless networking there have been two types of wireless networks: the **infrastructure** network, including some local area networks (LANs), and the ad hoc network. Ad hoc is Latin meaning "for this purpose." Ad hoc networks therefore refer to networks created for a particular purpose. They are often created **on-the-fly** and for one-time or temporary use. Often, ad hoc networks **are comprised** of a group of workstations or other wireless devices which communicate directly with each other to exchange information. Think of these connections as **spontaneous** networks, available to whoever is in a given area.

An ad hoc network is one where there are no access points passing information between participants. Infrastructure networks pass information through a central information hub which can be a hardware device or software on a computer. Office networks, for example, generally use a server to which company workstations connect to receive their information. Ad hoc networks, on the other hand, do not go through a central information hub.

Ad hoc networks are generally closed in that they do not connect to the Internet and are typically created between participants. But, if one of the participants has a connection to a public or private network, this connection can be shared among other members of the ad hoc network. This will allow other users on the spontaneous ad hoc network to connect to the Internet as well.

Ad hoc networks are common for **portable** video game systems like the Sony PSP or the Nintendo DS because they allow players to link to each other to play video games wirelessly. Some retail stores even create networks within them to allow customers to obtain new game demos via the store's own ad hoc network.

An ad hoc network can be thought of as a peer-to-peer network for the wireless age. Peer-to-peer or workgroup style networks were used to create a network environment for early Windows computers. This allowed these early computers to connect to each other to exchange information, usually in a smaller office environment without the need for domains and the additional management and overhead that comes with them.

The possibilities with ad hoc networks are quite endless. With connection sharing and other methods to access public or private wired or wireless networks, the reach of an ad hoc network can extend quite far.

출처: http://www.wisegeek.com/what-is-an-ad-hoc-network.htm#il

이번 장에서 꼭 알아야 할 단어

inception

시작

Inception is the beginning of an organization or official activity.

- Since its **inception** in 1972, the company has been at the forefront of computer development.
- We take a look at funds that have outperformed since their **inception**.
- A decade after its **inception**, it had virtually ceased to function.

infrastructure

(사회) 기반 시설

*An **infrastructure** is the basic systems and services, such as transport and power supplies, that a country or organization uses in order to work efficiently.*

- The war has badly damaged the country's **infrastructure.**
- The New Zealand Government has announced a new, national approach to planning, funding and implementing the nation's future **infrastructure** needs.

on-the-fly

진행(구동) 중에 이루어지는

*In relation to computer technology, "**on the fly**" describes activities that develop or occur dynamically rather than as the result of something that is statically predefined.*

***On-the-fly programming** is a style of programming in which the programmer/performer/composer augments and modifies the program while it is running, without stopping or restarting, in order to assert expressive, programmable control for performance, composition, and experimentation at run-time.*

be comprised of

~으로 구성되다

*If something **comprises** or **is comprised of** a number of things or people, it has them as its parts or members.*

- The class **is comprised** mainly **of** Italian and French students.
- The School of Computer Science in the College of Computing **is comprised of** faculty and students engaged in research and teaching within computing systems, broadly defined, and computing theory.
- Regular computers **are** generally **comprised of** a housing for mounting a computer mother board. According to the prior art, the computer housing is specifically designed according to the type of computer.

spontaneous

즉흥적인

***Spontaneous** acts are not planned or arranged, but are done because someone suddenly wants to do them.*

- His jokes seemed **spontaneous,** but were in fact carefully prepared beforehand.
- She's such a **spontaneous,** lively woman.

portable

휴대(이동)가 쉬운

*A **protable** machine or device is designed to be easily carried or moved. The machine or device can be called a portable.*

- A **portable** application (portable app) is a computer software program that is able to run independently without the need to install files to the system it is run upon.
- We can show you where to find the best **portable** products of all types through our comprehensive, search-friendly indexes.
- You can't take that on our trip because it isn't **portable.**

Common Collocations

Inception	Infrastructure
Inception cohort Inception report Inception phase	Business infrastructure Infrastructure funds Network infrastructure Technology infrastructure
on-the-fly	**portable**
Think on-the-fly learning on-the-fly cocktails on-the-fly sport fishing on-the-fly	portable software portable dvd players portable games portable generator

6.2 Useful Expressions

go through : ~을 거치다, ~을 겪다/경험하다	
• I went through a lot last year with my divorce. • After you go through difficulties, you come out stronger. • After going through customs, you can meet your family.	• 나는 작년에 이혼하면서 많은 것을 겪었다. • 어려움을 겪고 난 후에는, 당신은 더 강해질 것이다. • 세관을 통과한 후에, 너는 가족을 만날 수 있다.
allow sb to V : sb가 V 하도록 (허용)하다.	
• Smoking is not allowed in this restaurant. • Our software will allow you to convert pdf files to Word documents. • This will allow our customers to replace their mobile calls with cheaper fixed-line calls.	• 이 식당에서 흡연은 허용되지 않는다. • 우리 소프트웨어는 당신이 pdf파일을 워드 문서로 바꾸는 것을 가능하게 할 것 이다. • 이것은 우리의 고객들이 그들의 무선통화를 더 저렴한 유선통화로 대체할 수 있게 할 것이다.

6.3 Descriptive Grammar

Since the inception of wireless networking there have been two types of wireless networks: the infrastructure network, including some local area networks (LANs), and the ad hoc network.

Colon (:) 은 목록(list)을 만들거나, 동격(appositive)을 나타내거나, 인용(quotation)을 하기 위해 사용된다. 먼저 목록을 나열하는 경우 콜론이 사용된 예를 살펴보도록 하자.

The daily routine should include at least the following : twenty knee bends, fifty sit-ups, fifteen leg lifts, and five minutes of running in place.
(하루의 일과는 최소한 다음의 것을 포함해야만 한다 무릎굽혀펴기 20번, 윗몸 일으키기 50번, 다리 올리기 15번 그리고 달리기 5분)

The following foods are prohibited : candy, gum, and chocolate.
(다음 음식들은 금지된다: 사탕, 껌, 그리고 초콜릿)

Let's invite the following people : Marcy, Jim, Katrina, and Sammy.
(다음 사람들을 초대하자: 마시, 짐, 카트리나, 그리고 새미

다음은 콜론이 동격을 나타내는 용도로 사용된 경우이다.
My roommate is guilty of two of the seven deadly sins: gluttony and sloth.
(내 룸메이트는 7대 죄악 중 두 개가 유죄이다: 폭식과 나태.)

아래 예문은 인용하는 문장 앞에 콜론을 사용한 경우이다.
Consider the words of John F. Kennedy: "Ask not what your country can do for you; ask what you can do for your country."
(John F. Kennedy의 말을 생각해 봐라: "국가가 당신을 위해 무엇을 할 수 있는 가를 묻지 말고 ; 당신이 국가를 위해 무엇을 할 수 있는지 물어라."

위의 문장을 다시 살펴보면,

Since the inception of wireless networking / there have been two types of wireless networks : the infrastructure network, including some local area networks (LANs), and the ad hoc network.
무선 네트워킹가 사용된 이후에 / 무선네트워크는 두 가지 유형이 존재해 왔다 : 몇몇 지역 네트워크를포함하는 Infrastructure 네트워크, 그리고 ad-hoc 네트워크

6.4 Substance

우선 무선 네트워크를 토폴로지 (Topology)[2]로 구분하면 2가지 방법으로 구분할 수 있다. 첫 번째는 인프라기반 네트워크이고, 나머지 하나는 Ad-hoc 네트워크이다.
인프라기반 네트워크는 말 그대로 기반 인프라을 이용하여 무선 네트워크를 구성하는 것을 의미한다.

영어 작문 연습

- To begin with, if we divide networks according to Topology, we can divide them two by methods. The one includes infrastructure networks; the other is an Ad-hoc networks. The infrastructure network means to constitute wireless network by using base infra as it is.

그럼, Ad-Hoc 네트워크라는 것은 무엇일까?
사람들은 개인적, 혹은 직업적 필요에 의해, 노트북 컴퓨터나 휴대폰, PDA, MP3 플레이어와 같은 휴대용 기기들을 들고 다닌다. 대체로 이들 기기들은 독립적으로 사용된다. 그런데 만약 직접 연동이 된다면 어떨까?

영어 작문 연습

- Then, what is an Ad-hoc network?
  People carry portable devices like notebook computers, cell phones, PDAs, and MP3 players. On the whole, these devices are used independently. But how about if these are interlocked with each other directly?

통신기기간 특별한 무선 통신을 'ad hoc 네트워크'라고 하는데, Ad hoc 네트워크는 기반 인프라가 없다. 즉, 중앙 시스템의 도움 없이 노드들에 의해 자율적으로 언제, 어디서나 기기간 통신을 자체적으로 라우팅하여 연결하는 무선 네트워크 토폴로지이다. 가정에서도 가전 기기들이 블루투스와 같은 통신 모듈만 내장되어 있다면, 무선 네트워크 구축은 간단하게 된다.

---

2 토폴로지란 네트워킹 구조 내에서의 네트워크 노드와 미디어의 물리적 구성이다

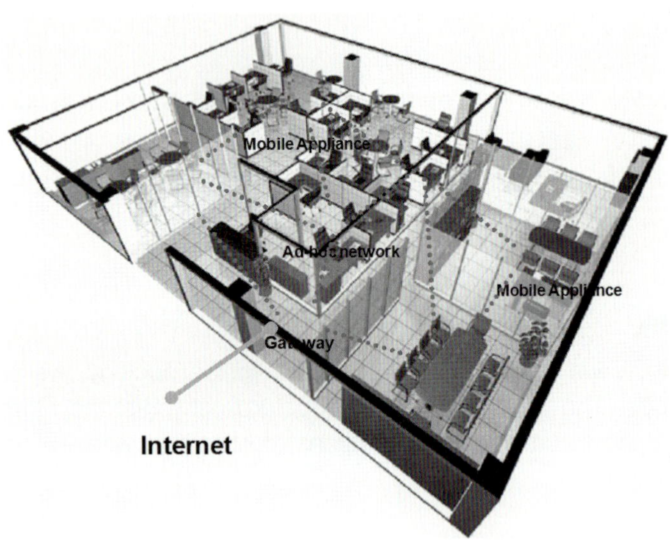

[그림 10] Ad-Hoc 네트워크 개념도

Ad-Hoc 네트워크는 독립적으로 존재할 수도 있고, 기간 망과 연동하는 경우가 있는데, 기간 망에 연결하고자 할때는 위의 그림처럼 게이트웨이를 통해 연결하면 된다.

영어 작문 연습

- An Ad-Hoc network can remain independent or be interlocked with an infra network. The Ad-Hoc interlocked with an infra network is linked to a gateway like in Figure5.

6.5 Glossary

- List of ad-hoc routing protocols
- Delay-tolerant networking
- Wireless community network
- Mesh network
- Wireless mesh network
- Wizzy Digital Courier
- Real Time Locating

References [참고문헌]

[1] 융합 그리고 넥스트, 손병희 저, 교우사, 2015
[2] Cambridge Advanced Learner's Dictionary, Cambridge University Press
[3] Collins COBUILD Intermediate Learner's Dictionary, Heinle & Heinle Publishers
[4] Longman Essential Activator, Longman National Corpus
[5] English Grammar in Use, Raymond Murphy, Cambridge University Press

6과 연습문제

01 Match each word in the first column with its definition in the second column.

1.	inception	A.	the beginning of an organization or official activity
2.	infrastructure	B.	describes activities that develop or occur dynamically
3.	on-the-fly	C.	consist of
4.	be comprised of	D.	the basic systems and services
5.	spontaneous	E.	easily carried or moved
6.	portable	F.	not planned

02 Each of the following excerpts is about ad hoc networks. Read the sentences. Fill in the blanks using the words in the below box.

1) A mobile ad hoc ________________(MANET), sometimes called a mobile mesh network, is a self-________________network of mobile devices connected by wireless ________________.

links, configuring, network

2) So what is a mesh network? Think of the electronic equivalent of party guests passing on the message that a child is missing until that child is found – many short range communications, in many directions – like a ________________– add up to finding the child further away. All those people were able to pass on a message in an ad hoc network that dissolved when its task is done. In the electronic equivalent, ________________can form ad hoc networks where many small ranges add up to a big one thus permitting the tags to be much

smaller, lower cost and more reliable than would otherwise be the case. However, here the ________________ operates without human intervention. It performs its tasks silently and automatically.

network, mesh, WSN

2) An ad-hoc network is a local area network (________________) that is built spontaneously as devices connect. Instead of relying on a base station to ________________ the flow of messages to each node in the network, the individual network nodes forward packets to and from each other. In Latin, ________________ literally means "for this," meaning "for this special purpose" and also, by extension, improvised or impromptu.

LAN, ad hoc, coordinate

3) On ________________ computer networks, ad-hoc mode is a method for wireless devices to directly communicate with each other. Operating in ________________ mode allows all wireless devices within range of each other to discover and communicate in peer-to-peer fashion without involving central access ________________ (including those built in to broadband wireless routers).

points, wireless, ad hoc

4) Ad hoc ________________ might seem like a much smarter alternative to using a wireless router, but it does have a couple of disadvantages: If the computer connected to the Internet is shut down, all computers that are part of the ad hoc network lose their Internet access. To connect to the ________________, one computer always needs a wired network ________________.

networking, Internet, connection

5) On networks with wireless ________________, the router has the important job of forwarding communications from the computers on your home network to the Internet. On ad hoc networks, you must designate one ________________to serve this role. The computer you choose must have a wired connection to the ________________, and it should be left on whenever you want to be able to use your other computers.

Internet, routers, computer

6) I've received many inquiries from Windows XP residential users asking how to wirelessly ________________a second or third computer without the expense of a wireless router or a router and a stand-alone wireless access point (WAP). This is sometimes called an ________________wireless network. Building an ad hoc 802.11b wireless network using the graphical user interface in Windows XP takes just a few minutes. Because it's powered by wireless zero configuration, you won't need to run a single cable or hunt for an electrical outlet! Add Internet Connection Sharing on the host computer and all ________________computers will be surfing the net wirelessly in no time at all.

connected, ad hoc, network

7) ________________Networks are self organizing, self healing, distributed ________________which most often employ wireless ________________techniques.

transmission, ad hoc, networks

8) Ad hoc networks, which cover a variety of network ________________for specific purposes, such as mobile ad hoc networks, ________________networks, vehicular networks, underwater networks, underground networks, local area networks, personal area networks,

home _______________, etc., promise a broad range of applications in civilian, commercial, and military areas.

networks, paradigms, sensor

9) Smart Ad Hoc Networks are capable of analyzing the radio propagation _______________ they operate in to optimize performance. This typically requires that the _______________ nodes have positioning capability as well as memory to recall _______________ local condition.

geographically, environment, network

7

Telematics

7 Telematics

7.1 Introduction

Telematics is a compound word of Telecommunication and **Informatics.** As the meaning of the word, it is the technology supplying safe and useful service such as an emergency service, a road guidance, a traffic report, etc. It may call devices or services supplying entertaining information like internet, movie, and games **collectively.**

This technology reconstruct vehicle to the third information living space (Connected Vehicle) following office and home by wired or wireless communication and broadcasting network.

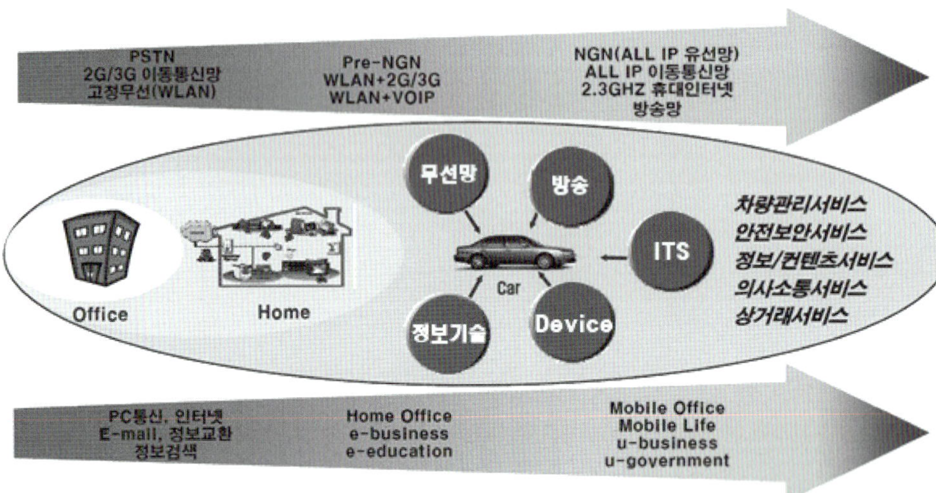

The term telematics (synonymous with telemetry) is used in a number of ways: More specifically it is the science of sending, receiving and storing information **via** telecommunication devices.

More recently, telematics have been applied specifically to the use of Global Positioning System technology integrated with computers and mobile communications technology in automotive navigation systems.

Most narrowly, the term has evolved to refer to the use of such systems within road vehicles, in which case the term vehicle telematics may be used.

- From Wikipedia, the free encyclopedia

이번 장에서 꼭 알아야 할 단어

Informatics
(=Information Science)
정보학

Informatics *Informatics is the study of information and the ways information is used by and affects human beings and social systems. Also it is a broad academic field including artificial intelligence, cognitive science, computer science, information science, and social science.*

☞ ***정보학***이란 정보 기술과 비슷하게 사용되는 용어로, 컴퓨터를 이용한 정보 조직 시스템 및 정보의 발생, 전달, 수집, 축적, 처리 등과 관련된 이론과 실제 운용에 관한 학문을 말한다.

collectively

집합적으로, 총괄적으로

Collective means shared by or involving every member of a group of people. Collectively is an adverb form.

- These men have **collectively** given over 600 years' service.
- It is an issue for all of us individually as well as **collectively.**
- She has a staff of four who **collectively** earn almost $200,000.

via

~을 통하여

If you do something via a particular means or person, you do it by making use of that means or person.

- She sent us the photo **via** e-mail.
- The London-Thailand flight goes **via** Korea.
- I only found out about it **via** my brother.

term

용어

A term is a word or expression with a specific meaning.

- The **term** we use now is 'assisted' arranged marriage
- Who was the first person to coin the **term** "the Internet"?
- 'Register' is the **term** commonly used to describe different levels of formality in language.

Common Collocations

Informatics	via
Nursing informatics Healthcare informatics Clinical informatics	via e-mail via + 지역이름 (Seoul / London/ Paris, etc) via satellite

7.2 Useful Expressions

be integrated with ~과 통합되다	
• Science is integrated with math in this class. • Non-competitive universities will be integrated with competitive universities. • Boys will be integrated with girls in this school starting from next year.	• 이 수업에서 과학은 수학과 통합된다. • 경쟁력이 없는 대학은 경쟁력이 있는 대학과 통합될 것이다. • 이 학교에서는 내년부터 소년들과 소녀들은 통합학급으로 운영될 것이다.
refer to: ~을 가리키다, 나타내다	
• Figures in parenthesis refer to page numbers. • Fans and media use it to refer to a top goal-scorer • They dress as women and refer to themselves as women	• 괄호 속에 있는 숫자들은 페이지를 가리킨다. • 팬들과 매체는 그걸 득점선두자를 나타내기 위해 사용해. • 그들은 여장을 하고 자신들을 여성이라고 부른다.

7.3 Descriptive Grammar

The etymology of Telematics, *as determined* by Automotive Telematics author and academic Dennis Foy, is from the Greek "tele" and . . . (이 문장이 본문 어디에? 확인해보기)

'as determined…' 이하는 분사구문으로 주절을 수식하는 부사구의 역할을 한다. 분사구문에서는 접속사가 생략되는 것이 일반적이나, 의미를 명확히 나타내기 위해서 접속사를 사용하기도 한다.
수동적 동작이나 상태를 나타내는 수동형 분사구문의 형태는 'being+과거분사' 이며 'being'이 생략되기도 한다.

As he was tired from the work, he went to bed earlier.
(그는 일에 지쳤으므로, 더 일찍 잤다.)
Tired from the work, he went to bed earlier.
(그는 일에 지쳤으므로, 더 일찍 잤다.)

- 분사구문 : 동사+~ing, 주어+동사: ~하면서, ~하다

Feeling confident, I sang the song.
(자신감을 느끼면서 나는 그 노래를 불렀다.)

위의 문장을 다시 살펴보면,

The etymology of Telematics, as determined by Automotive Telematics author and academic Dennis Foy, is from the Greek "tele" and…
텔레매틱스(Telematics)의 어원은, Automotive Telematics author and academic Dennis Foy에 의해 정해졌듯이, 그리스의 이야기로부터 왔으며…

7.4 Substance

이종 통신 방송망과 지능형 단말기를 통해 홈 네트워크, 사무자동화 등과 연계함으로써 가정과 사무실에 이어 다양한 유비쿼터스 서비스를 차량에서도 Seamless Service[3]를 제공받을 수 있게 되었다. 제공하는 서비스틑 차량관리 서비스, 안전보안 서비스, 정보 및 컨텐츠 서비스, 의사소통 서비스, 상거래 서비스 등이 있다.
차량이라는 제한적인 공간에서 유비쿼터스 환경에 적합한 융합서비스를 실현한 것이다. 뿐만 아니라, 여기에 ITS(Intelligent Transport System)[4]와 접목시켜 도로 및 교통관리, 교통정보를 제공한다.

영어 작문 연습

- Linked to home network and office automation through the mobile communication, TV network and intelligent devices, in also the vehicle various ubiquitous service (Seamless Service) can be offered like in home and office. Those services are car care service, safety guard service, information and contents service, communication service, and commercial transaction service. In the limitary space as vehicle the integrated service of Ubiquitous environment was realized. Also, with engrafted ITS (Intelligent Transport System) it offers road and traffic control and traffic reports.

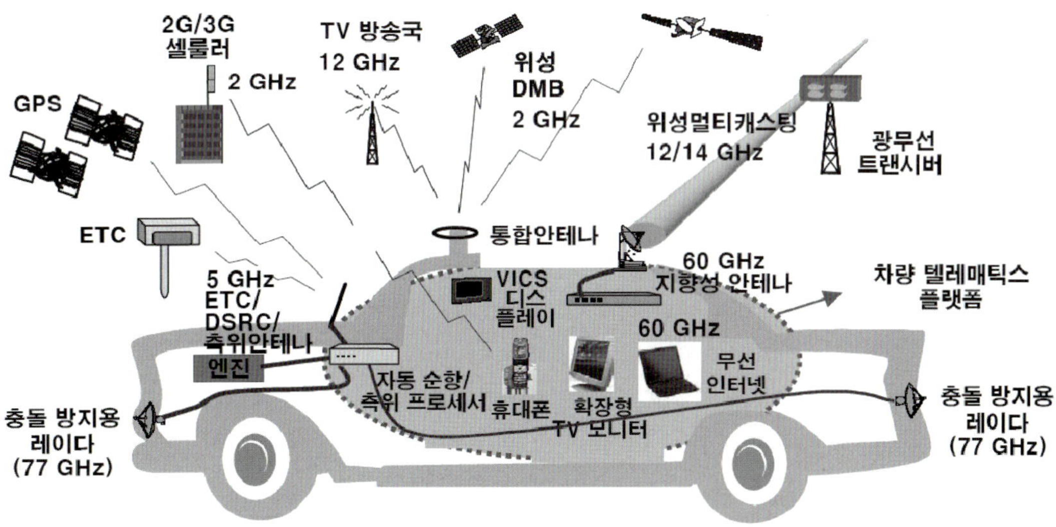

3 사용자가 어떤 어플리케이션을 사용할 때 그 어플리케이션이 네트워크에 끊김 없이 연결되는 서비스를 말한다.
4 전자, 정보, 통신, 제어 등의 기술을 교통체계에 접목시킨 지능형 교통 시스템이다.

ETC (Electronic Toll Collection), DSRC (Dedicated Short Range Communications), DMB (Digital Multimedia Broadcasting), VICS (Vehicle Information and Communication System)

– From http://dwcij.com.ne.kr/newtech/new_jpnits.htm

7.5 Glossary

- Artificial Passenger
- Automotive navigation system
- Carputer
- Floating car data
- GNSS Road Pricing
- Information communication technology
- Infotainment
- Intelligent transportation systems
- Intelligent vehicle technologies
- Map database management
- Mass surveillance
- Telematics for Libraries Program
- Telemetry

References [참고문헌]

[1] 융합 그리고 넥스트, 손병희 저, 교우사, 2015

[2] Cambridge Advanced Learner's Dictionary, Cambridge University Press

[3] Collins COBUILD Intermediate Learner's Dictionary, Heinle & Heinle Publishers

[4] Longman Essential Activator, Longman National Corpus

[5] English Grammar in Use, Raymond Murphy, Cambridge University Press

7과 연습문제

01 Match each word in the first column with its definition in the second column.

1. collectively	A. a word or expression
2. via	B. by making use of that means or person
3. term	C. shared by every member of a group of people
4. refer to	D. mention
5. be integrated with	E. be combined with

02 Each of the following excerpts is about telematics. Read the sentences. Fill in the blanks using the words in the box.

1) ________________ is the convergence of telecommunications and information ________________, the term later evolved to refer to automation in automobiles, such as the invention of the emergency warning system for vehicles. GPS navigation, integrated hands-free cell phones, wireless safety communications and ________________ driving assistance systems all are covered under the telematics umbrella.

processing, telematics, automatic

2) Telematics ________________ are self-orientating open network architecture structure of variable programmable intelligent beacons developed for application in the development of intelligent vehicles — with target intent to accord (blend, or mesh) warning ________________ with surrounding vehicles in the vicinity of

travel, intra-vehicle, and infrastructure. Emergency warning ________________ for vehicles telematics particularly developed for international harmonization and standardization of vehicle-to-vehicle — infrastructure-to-vehicle — and vehicle-to-infrastructure real-time Dedicated Short Range Communication (DSRC) systems.

technologies, system, information

3) Telematics is the ________________ of computers and wireless telecommunications technologies, ostensibly with the goal of efficiently conveying ________________ over vast networks to improve a host of business functions or government-related public services. The most notable example of ________________ may be the Internet itself, since it depends on a number of computer networks connected globally through telecommunication backbones.

information, blending, telematics

4) Cold store freight trailers that are used to deliver fresh or frozen foods are increasingly incorporating ________________ to gather time-series data on the temperature inside the cargo container, both to trigger alarms and record an audit trail for business purposes. An increasingly sophisticated array of ________________, many incorporating ________________ technology, are being used to ensure that temperature throughout the cargo remains within food-safety parameters.

telematics, RFID, sensors

5) Major automakers are equipping new prototype vehicles with wireless-based services controlled by voice commands. This kind of ________________ could enable motorists to perform a variety of wireless functions such as accessing the Internet, receiving or sending e-mail, downloading digital audio and video ________________, or obtaining

"smart" transportation ____________________.

files, telematics, information

6) Wireless vehicle safety communications ____________________ aid in car safety and road safety. It is an electronic sub-system in a car or other vehicle for the purpose of exchanging safety ____________________, about such things as road hazards and the locations and speeds of vehicles, over short range radio links. This may involve temporary ad hoc wireless local area ____________________.

networks, information, telematics

7) The ____________________ industry is not limited to automotive applications. Other applications are being studied or developed for monitoring water and air pollution, for medical informatics and health care, and for distance learning. Many European countries are ____________________ uniform policies to integrate telematics ____________________ into government, business and education.

applications, developing, telematics

8) ____________________ is now a somewhat dated term, but has been used to describe the convergent action of computer and communication (terrestrial and satellite) technologies. Another related term is ICT – Information and Communications ____________________ – essentially the same thing. In other words, telematics is about the merging of computers and telecommunications. Telematic technologies include satellite television, the use of computers to mediate ____________________ known as CMC, visual collaboration through the use of video conferencing and of course the Internet.

telematics, communication, technologies

9) With the current financial constraints and the increasing pressure on education providers to perform better, ________________ offers a viable solution for sustainability and growth. Widening catchment areas and increasing student numbers without ________________ access to new resources such as more classrooms and staff can be achieved using telematics. There is also the move towards lifelong ________________ to consider.

telematics, learning, direct

10) Telematics comprise electronic, electromechanical, and electromagnetic ________________ — usually silicon micromachined components operating in conjunction with ________________ controlled devices and radio transceivers to provide precision repeatability functions (such as in robotics artificial ________________ systems) emergency warning validation performance reconstruction.

intelligence, devices, computer

8

Embedded Systems

8 Embedded Systems

8.1 Introduction

An **embedded** system is a special-purpose computer system designed to perform one or a few dedicated functions, often with real-time computing **constraints**. It is usually embedded as part of a complete device including hardware and mechanical parts. In contrast, a general-purpose computer, such as a personal computer, can do many different tasks depending on programming. Embedded systems control many of the common devices in use today.

Since the embedded system is dedicated to specific tasks, design engineers can **optimize** it, reducing the size and cost of the product, or increasing the reliability and performance. Some embedded systems are mass-produced, benefiting from economies of **scale**.

Physically, embedded systems range from portable devices such as digital watches and MP4 players, to large stationary installations like traffic lights, factory controllers, or the systems controlling nuclear power plants. Complexity varies from low, with a single microcontroller chip, to very high with multiple units, **peripherals** and networks mounted inside a large chassis or enclosure.

In general, "embedded system" is not an exactly defined term, as many systems have some element of programmability. For example, Handheld computers share some elements with embedded systems — such as the operating systems and microprocessors which power them — but are not truly embedded systems, because they allow different **applications** to be loaded and peripherals to be connected.

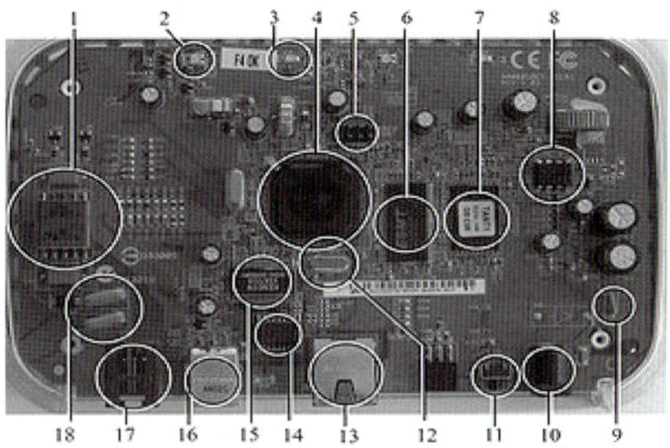

Picture of the internals of a Netgear ADSL modem/router. A modern example of an embedded system. Labelled parts include a microprocessor (4), RAM (6), and flash memory (7).

- *From Wikipedia, the free encyclopedia*

이번 장에서 꼭 알아야 할 단어

embedded

내장된

*If an object is **embedded** in something, it is fixed there firmly and deeply.*

- There is glass **embedded** in the cut in his leg.
- I think that hatred of the other is deeply **embedded** in our society.

☞ embedded system이란 특정한 제품이나 솔루션에서 주어진 작업을 수행할 수 있도록 추가로 탑재되는 솔루션이나 시스템을 말한다.

constraint

제약, 제한

*A **constraint** is something which controls what you do by keeping you within particular limits.*

- The **constraints** of politeness wouldn't allow her to say what she really thought about his cooking.
- Financial **constraints** on the company are preventing them from employing new staff.
- She tried to appear friendly, but her **constraint** was obvious.

optimize

~을 최적화하다

*To **optimize** is to make something as good as possible.*

- We need to **optimize** our use of the existing technology.
- I need an overview of why and how one can **optimize** code for speed.

scale

규모, 범위

***Scale** is the size or level of something.*

- We don't yet know the **scale** of the problem.
- Nuclear weapons cause destruction on a massive **scale.**
- My parents used to entertain friends on a large **scale.**

peripheral

(컴퓨터의) 주변 장치

*A **peripheral** is a piece of equipment, such as a printer, that can be connected to a computer.*

- The operating system contains all the drivers for the **peripherals** attached to the computer.
- The part of a machine instruction that references data or a **peripheral** device.

  ☞ *(컴퓨터) 주변 장치 (중앙 처리 장치와 대비해 카드 천공기·라인 프린터·자기 테이프 장치 등)*

application

(컴퓨터) 응용 프로그램

*An **application** is a computer program that is designed for a particular purpose.*

- Many software companies and internet providers believe that e-learning is the next killer **application.**
- This site is basically a showcase of possible projects one could do using various computer **applications.**
- **Application software** is computer software designed to help the user perform a particular task.

Common Collocations

Embed	Optimize(optimization)	Scale
Embed music Embed games Embed flash Embed tag	Optimize Vista Scholastic optimization Mathematical optimization Linear optimization Continuous optimization	Musical scale Map scale Gram scale Plant scale Weight scale
Peripheral	**Mounted**	**Enclosure**
Computer peripheral Peripheral neuropathy Peripheral nervous system Peripheral devices Peripheral nerve	Mounted games Mounted addon Mounted infantry Mounted drive	Enclosure movement Electrical enclosure Electronic enclosure

8.2 Useful Expressions

be dedicated to : ~에 헌신하다. 전념하다	
• She's completely dedicated to her work.	• 그녀는 그녀의 일에 완전히 헌신했다.
• The Party is dedicated to protecting the environment.	• 그 파티는 환경을 보호하기 위해 열렸다.
• Other university presidents have been dedicated to drawing funds to their schools.	• 다른 대학의 총장들은 학교에 자금을 유치하는데 전념해 오고 있다.
portable device : 휴대용 장치	
• A lighter is a portable device used to create a flame.	• 라이터는 불을 만들기 위해 사용되는 휴대용 장치이다.
• An iPod is a portable device for listening to music and playing games.	• 아이패드는 음악을 듣고 게임을 할 수 있는 휴대용 장치이다.
• I have been working on gathering more information about a new portable device.	• 나는 새로운 휴대용 장치에 대해 더 많은 정보를 모으기 위해 일해왔다.

8.3 Descriptive Grammar

In general, "embedded system" is not an exactly defined term, *as* many systems have some element of programmability.

"as + 주어 + 동사" 형태로 쓰인 경우 접속사로써 문맥에 따라 4가지 의미가 있다.

As you leave the train, watch your step. [when : ~할 때에]
기차에서 내릴 때, 조심해라.
As it was cold in the morning, I caught cold. [Because : ~이기 때문에]
오늘 아침에 추웠기 때문에, 나는 감기에 걸렸다.
Poor as he was, he had won the general respect of the neighborhood.
[Although : 비록 ~일지라도] 비록 가난했지만, 그는 이웃들에게 존경을 받았다.
☞ As poor as he is에서 앞의 as가 'poor'를 강조하기 위해 생략된 형태임에 유의하자.
As you grow older, you will come to know the world. [~함에 따라]
점점 나이가 들어감에 따라, 당신은 세상을 알게 될 것이다.

위의 문장을 다시 살펴보면,

In general, "embedded system" is not an exactly defined term, as many systems have some element of programmability.
일반적으로, "임베디드 시스템"은 정확하게 정의된 용어가 아니다, 많은 시스템들이 프로그램작동 가능한 요소를 가지고 있기 때문에
☞ (2)의 의미로 사용된 것임을 알 수 있다.

8.4 Substance

일반적으로 시스템의 정의를 내려 보면, 입력을 처리해서 출력을 만들어 내는 것을 말하며 시스템 구성도는 다음과 같다.

영어 작문 연습

- The general definition of a system is that which processes inputs in order to make outputs. The configuration of a system is as follows.

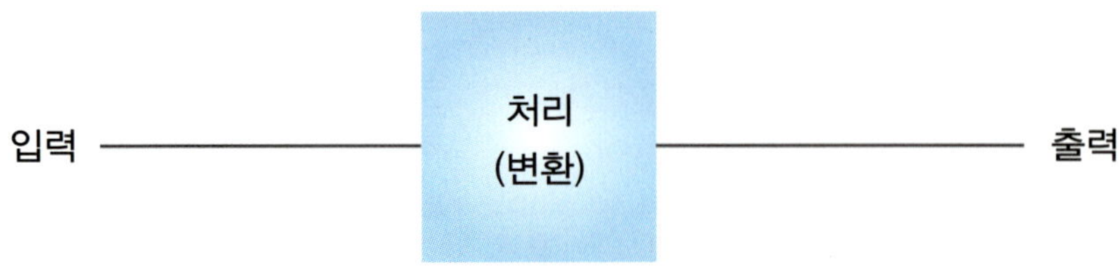

[그림 11] 시스템 구성도

이러한 관점에서 보면 사람도 시스템이다. 입력과 출력이 있기 때문이다. 그러나 시스템이라고 하면 보통 컴퓨터를 연상하게 되는데, 이것은 아마도 컴퓨터가 데이터를 전문적으로 처리하는 기계이기 때문일 것이다. 좀 더 수학적으로 표현하면 시스템은 입력 데이터를 출력 데이터로 변환하는 것으로 볼 수 있다.

영어 작문 연습

- From this point of view, a human is also a system. For it do inputs and outputs. But if we say system, we recall computers usually, because computer is the machine processing data professionally. If we express it more mathematically, a system is the thing converting input data to output data.

그럼, 임베디드 시스템이란 무엇일까?

임베디드란 어느 시스템에 내장된 (Embedded) 시스템으로서 마이크로프로세서 혹은 마이크로컨트롤러를 내장하여 원래 제작자가 의도했던 특정 기능만을 수행하도록 제작된 컴퓨팅 장치라고 한다. 내장형 시스템이기 때문에 소형이 많지만, 모든 시스템이 다 소형인 것은 아니다. 응용 분야를 보더라도 PC와 같은 범용 컴퓨팅 시스템을 제외한, 특정 기능만을 수행하는 모든 컴퓨팅 시스템이 임베디드 시스템이므로 다양한 산업 분야의 기기들이 포함

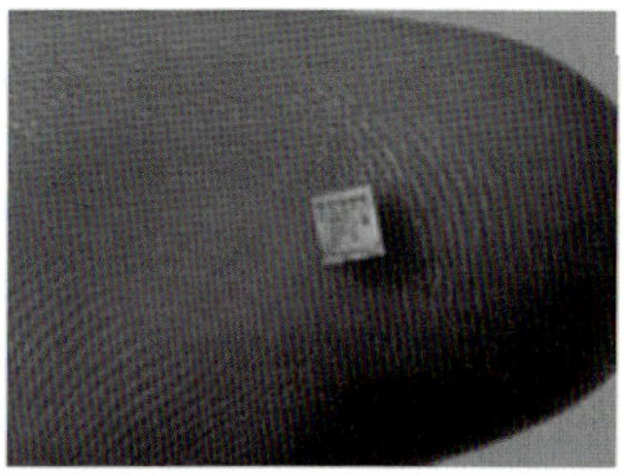

된다. 따라서, 산업 분야의 스펙트럼도 정보 가전에서 시작해서 정보 단말 등은 물론이고 산업/제어, 로봇, 사무자동화, 빌딩자동화, 산업 자동화, 군사, 통신, 물류/금융, 자동차/운송장비, 의료, 게임, 항공 관제 등 폭넓고 다양하게 응용되고 있는 것이다.

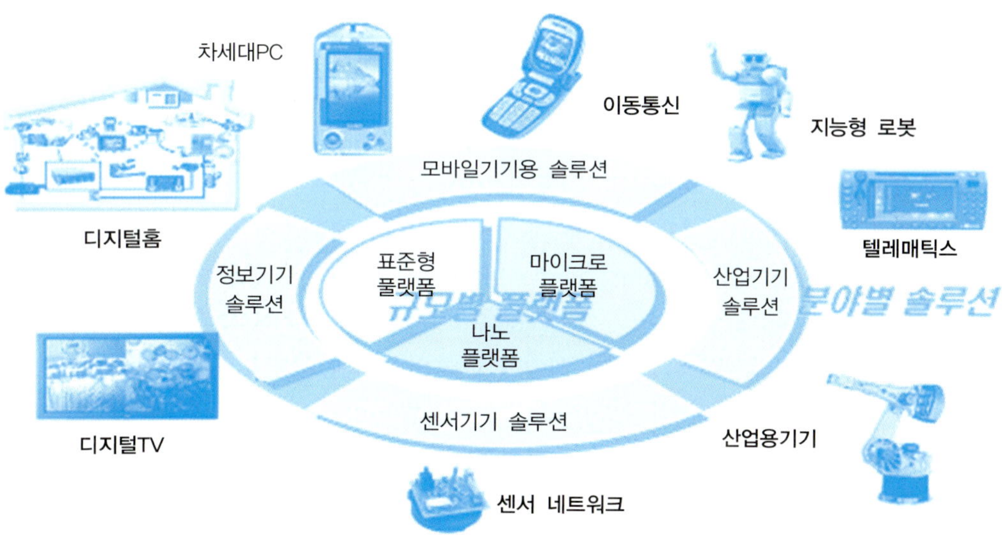

[그림 12] 임베디드 시스템 응용 분야

8.5 Glossary

- Communications server
- Cyber-physical system
- DSP
- Electronic Control Unit
- Embedded Hypervisor
- Embedded operating systems
- Embedded software
- Firmware
- Information appliance
- Microprocessor
- Programming languages
- Real-time operating system
- Software engineering
- System on a chip
- System on module
- Ubiquitous computing

References [참고문헌]

[1] 융합 그리고 넥스트, 손병희 저, 교우사, 2015
[2] Cambridge Advanced Learner's Dictionary, Cambridge University Press
[3] Collins COBUILD Intermediate Learner's Dictionary, Heinle & Heinle Publishers
[4] Longman Essential Activator, Longman National Corpus
[5] English Grammar in Use, Raymond Murphy, Cambridge University Press

8과 연습문제

01 Match each word in the first column with its definition in the second column.

1. embedded	A. to make something as good as possible
2. constraint	B. the size or level of something
3. optimize	C. limitation
4. scale	D. fixed
5. peripheral	E. a computer program designed for a particular purpose.
6. application	F. a piece of equipment, such as a printer

02 Each of the following excerpts is about embedded systems. Read the sentences. Fill in the blanks using the words in the below box.

1) An embedded ______________ is a computer system designed to perform one or a few dedicated functions, often with real-time computing constraints. It is ______________ as part of a complete device often including hardware and mechanical parts. In contrast, a general-purpose computer, such as a personal computer, is designed to be flexible and to meet a wide range of an end-user's needs. Embedded systems control many of the common ______________ in use today.

system, devices, embedded

2) The challenge of programming an ______________ system is that it is difficult to get real-time feedback from the system without a display. It is common to use a simple serial interface for diagnostic purposes, for example by connecting to a PC running terminal software via a

RS-232 to USB adapter. Also, embedded ______________ usually have very strict memory limitations, processor limitations, and speed limitations that must play a factor in designing an embedded system, and ______________ an embedded computer.

embedded, programming, systems

3) Physically, embedded systems range from portable ______________ such as digital watches and MP3 players, to large stationary installations like traffic lights, factory controllers, or the ______________ controlling nuclear power plants. Complexity varies from low, with a single microcontroller chip, to very high with multiple ______________, peripherals and networks mounted inside a large chassis or enclosure.

units, devices, systems

4) ______________ systems evolved with general-purpose ______________ systems. They were at the forefront of the use and justification of computer ______________, the closest to the field. This is where the theories are making or breaking.

computer, products, embedded

5) In today's world, embedded ______________ are everywhere -- homes, offices, cars, factories, hospitals, plans and consumer ______________. Their huge numbers and new complexity call for a new design approach, one that emphasizes high-level tools and hardware/software tradeoffs, rather than low-level assembly-language programming and logic ______________.

design, electronics, systems

6) An embedded system is a specialized computer ________ that is part of a larger system or machine. Typically, an embedded system is housed on a single microprocessor board with the ________ stored in ROM. Virtually all appliances that have a digital interface – watches, microwaves, VCRs, cars – utilize ________ systems. Some embedded systems include an operating system, but many are so specialized that the entire logic can be implemented as a single program.

embedded, system, programs

7) In the past, most of the embedded research and development was identified with real-time systems and industrial ________, but things have been changing since. With wider deployment of ________, the need for embedded systems has ________.

settings, increased, computers

8) In general, "embedded system" is not an exactly defined term, as many ________ have some element of programmability. For example, handheld ________ share some elements with embedded systems such as the operating systems and microprocessors which power them, but are not truly embedded systems, because they allow different ________ to be loaded and peripherals to be connected.

systems, applications, computers

9) An embedded system is some ________ of computer hardware and software, either fixed in capability or programmable, that is specifically designed for a particular function. Industrial ________, automobiles, medical equipment, cameras, household appliances, airplanes, vending machines and toys (as well as the more

obvious cellular phone and PDA) are among the myriad possible hosts of an _______________ system.

combination, embedded, machines

10) Over recent years, _______________ systems have gained an enormous amount of processing _______________ and functionality. Many of the formerly external components can now be integrated into a single System-on-Chip. This tendency has resulted in a dramatic reduction in the size and cost of embedded _______________. As a unique technology, the design of embedded systems is an essential element of many innovations.

embedded, systems, power

9

Robots

9 Robots

9.1 Introduction

A robot is a virtual or mechanical artificial agent. In practice, it is usually an electro-mechanical system which, by its appearance or movements, conveys a sense that it has intent or agency of its own. The word robot can refer to both physical robots and virtual software agents, but the latter are usually referred to as bots. There is no **consensus** on which machines qualify as robots, but there is general agreement among experts and the public that robots tend to do some or all of the following: move around, operate a mechanical limb, sense and manipulate their environment, and exhibit intelligent behavior, especially behavior which **mimics** humans or other animals.

Stories of artificial helpers and companions and attempts to create them have a long history but fully **autonomous** machines only appeared in the 20th century. The first digitally operated and programmable robot, the Unimate, was installed in 1961 to lift hot pieces of metal from a die casting machine and stack them. Today, commercial and industrial robots are in widespread use performing jobs more cheaply or with greater accuracy and reliability than humans. They are also employed for jobs which are too dirty, dangerous or dull to be suitable for humans. Robots are widely used in manufacturing, **assembly** and packing, transport, earth and space exploration, surgery, weaponry, laboratory research, and mass production of consumer and industrial goods [2].

People have a generally positive perception of the robots they actually encounter. Domestic robots for cleaning and maintenance are increasingly common in and around homes. There is anxiety, however, over the economic effect of automation and the threat of robotic weaponry, anxiety which is not helped by the **depiction** of many **villainous**, intelligent, **acrobatic** robots in popular entertainment. Compared with their fictional **counterparts**, real robots are still **benign**, **dim-witte**d, and clumsy.

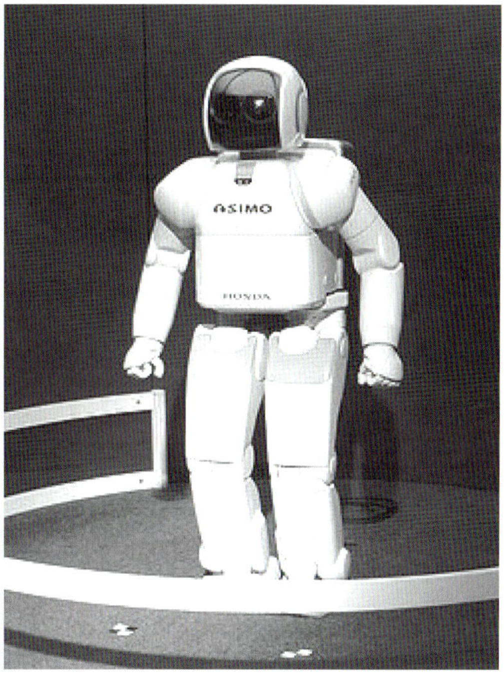

ASIMO, a **humanoid** robot manufactured by Honda.

- From Wikipedia, the free encyclopedia

이번 장에서 꼭 알아야 할 단어

consensus
의견일치, 합의

*A **consensus** is general agreement amongst a group of people.*
- There is a growing **consensus** in Europe that more action is needed.
- Could we reach a **consensus** on this matter? Let's take a vote.
- The general **consensus** in the office is that he's useless at his job.

mimic
~을 흉내내다, 모방하다

*If you **mimic** someone's actions or voice, you imitate the in an amusing or entertaining way.*
- He could **mimic** anybody.
- We like to get him to **mimic** our music teacher.
- She was **mimicking** the various people in our office.

autonomous
주체적인, 자주적인

= independent and having the power to make your own decisions
- They are no more free or **autonomous.**
- They proudly declared themselves part of a new **autonomous** province
- He treated us as **autonomous** individuals who had to learn to make up our own minds about important issues.

assembly
조립

***Assembly** is the process of putting together the parts of a machine or structure.*
- Putting the bookcase together should be a simple **assembly** job.
- The frame needs to be strong enough to support the engine **assembly.**
- For the rest of the day, he worked on the **assembly** of an explosive device.

depiction
묘사, 서술

***Depiction** is the way that something is represented or shown.*
- I don't like the **depiction** you are presenting of our company.
- The painter's **depictions** of the horror of war won her a worldwide reputation.
- I disapprove of the **depiction** of violence on television.

counterpart

상대, 대응관계

A ***counterpart*** *is a person or thing which has the same purpose as another one in a different place or organization.*

- The Prime Minister is to meet his European **counterparts** to discuss the war against drugs.
- He is her perfect **counterpart.**
- What is the **counterpart** for this system?

benign

상냥한, 유순한

You use benign to describe someone who is kind, gentle, and harmless.

- She is a **benign** old lady.
- They are normally a more **benign** audience.
- Though he talks in a rough manner, his actions are **benign.**

dim-witted

우둔한

When a message or electronic signal is ***transmitted,*** *it is sent by wires, radio waves, or satellite.*

- Marilyn was portrayed as some sort of **dim-witted**, floozy blonde.
- I hate to be approached by **dim-witted** salesman.
- ven a **dimwit** can use this software.

humanoid

휴머노이드 로봇

A ***humanoid*** *is a hybrid term from Latin humanus "human" and the Greek -oeides expressing likeness.*

☞ ***A humanoid robot*** *is a robot with its overall appearance based on that of the human body, allowing interaction with made-for-human tools.*

- That robot is so realistic that it is **humanoid.**

Common Collocations

Consensus	Autonomous
Consensus sequence Consensus decision making Consensus theory Consensus group	Autonomous system Autonomouos software Autonomous person Autonomous robot
Benign	**Humanoid**
Benign tumor Benign cancer Benign cyst Benign online	Humanoid fashion Humanoid robot Humanoid aliens Flying humanoid

9.2 Useful Expressions

in widespread use : 널리 이용되는	
• Because of high costs, hydrogen is not yet in widespread use. • Apple's iPods are in widespread use in America. • Four have been accused in widespread use of making fake money.	• 높은 비용 때문에, 수소는 아직 널리 이용되지 않는다 • 애플사의 ipod는 미국에서 널리 이용되고 있다. • 네 명은 위조화폐를 퍼뜨린 혐의를 받고 있다.
suitable for ∼에 적합한	
• Your dress is not suitable for the party. • Would now be a suitable moment for throwing a party? • The universally admitted that he is suitable for the job.	• 네 옷은 파티에 적합하지 않다. • 지금이 파티를 열기에 적합한 때일까요? • 대학측은 그가 그 일에 적합하다는 것을 인정했다.

9.3 Descriptive Grammar

People have a generally positive perception of the robots *they actually encounter*.

관계대명사[who/which/that]는 절[주어와 동사를 가지고 있는 문장]의 형태로 명사를 수식해 주는 역할을 한다. 명사를 꾸며준다는 점에서 문장에서 '형용사' 역할을 한다고 보면 되지만, 관계대명사는 명사 뒤에서 절의 형태로 길게 수식한다는 특성이 있다. 관계대명사를 포함한 절이 명사를 꾸며주는 경우를 살펴보자.

- the people [*who* live next door]
- the pictures [*which* you sent me]
- the movie [*that* I love]

예문을 보면 모두 관계사절 앞 명사(people, pictures, movie)에 대해 보충 설명을 하고 있다.

또한 관계사는 문장 내 어떤 역할을 하느냐에 따라 몇 가지로 나눠볼 수 있는데, 오늘은 문장 내 목적어 역할을 하는 경우를 살펴보도록 하자.

- Have you seen the pants *(which)* my boy friend bought for me?
- This is the movie *(that)* I've wanted to watch.

Which와 that 모두 관계사절 속에서 목적어 역할을 하고 있다. 즉, 'Have you seen the pictures?'와 'My boy friend bought *the pants* for me.' 이처럼 관계대명사가 목적어 역할을 할 때에는 관계대명사를 생략하는 것이 가능하다.

위의 문장을 다시 살펴보면,

People / have / a generally positive perception of the robots [they actually encounter].
사람들은 / 가지고 있다 / 일반적으로 긍정적인 생각을 / 로봇에 대한 / 그들이 실제로 만나는

9.4 Substance

로봇은 인공지능의 한 분야로 발전되어 왔다. 인공지능의 목적이 이성적으로 생각하는 시스템과 이성적으로 행동하는 시스템을 만드는 것이라면, 이 중 이성적으로 행동하는 시스템이 바로 로봇이다.

이런 로봇을 분류해 보면, 1999년도 일본에서는 산업용 로봇과 인간 공존형 로봇으로 분리했고, 미국에서는 서비스 개념을 강조하여 서비스에 따라 로봇으로 분류했다. 그 분류는 다음과 같다.

영어 작문 연습

- Robot has been developed as the part of artificial intelligent. If the purpose of artificial intelligent is to make a system able to think and a system able to act reasonably, the system able to act reasonably is robot.
- In Japan, the robot is classified industrial robot and coexistence type robot while in USA the robot is classified according to services emphasizing service concept. The classification is as follows.

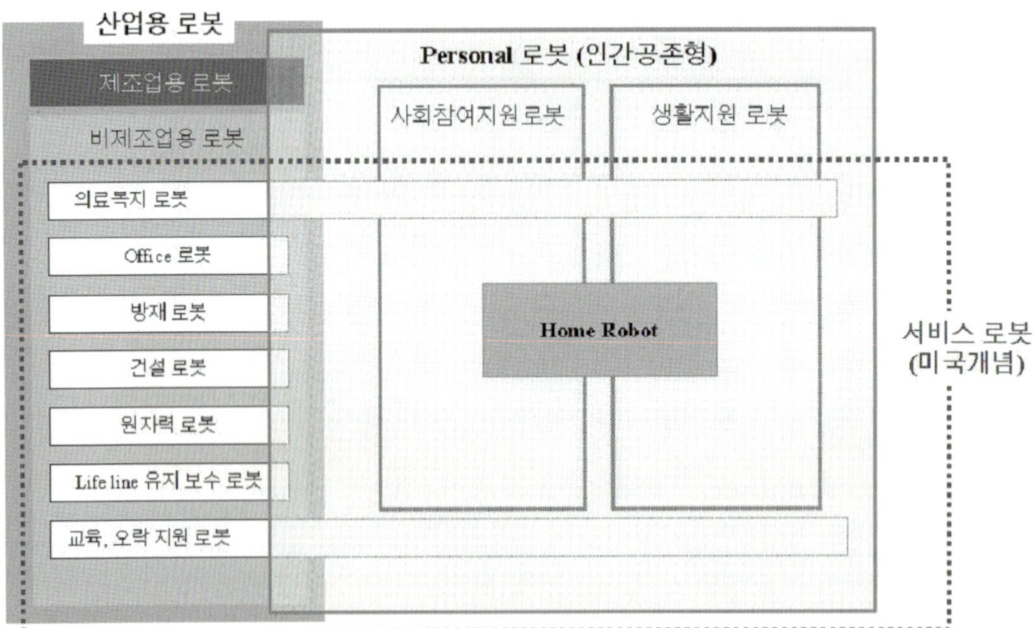

[그림 13] 로봇 분류

(자료 출처 : 일본로봇공업회, 1999)

그럼 로봇의 정의를 무엇일까? 로봇은 동물의 육체적 기능을 한가지 이상 수행할 수 있는 기계로 정의된다.

영어 작문 연습

- Then, what is the definition of a robot? A robot is defined as the machine which can accomplish at least one creature's physical function.

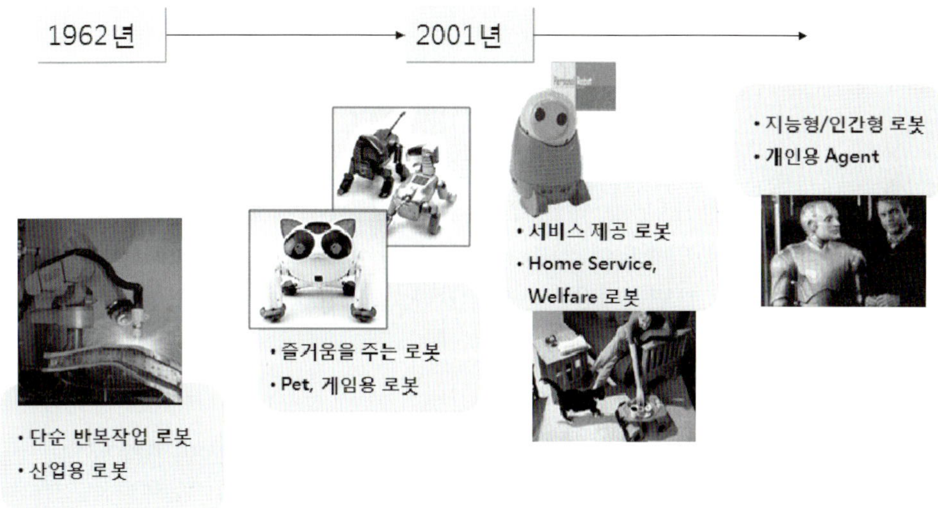

[그림 14] 로봇 발전 흐름

이 로봇은 1962년부터 산업용 로봇으로부터 출발하게 되는데, 이런 산업용 로봇은 주로 짐을 나르거나 용접하는 기능을 수행하였다. 2001년도부터는 지능형 로봇이라고 해서 자신의 목적을 가지고 있으며 입력된 정보에 따라 행동을 바꿀 수 있는 로봇으로, 물리적 센서에 의해 입력 정보를 수집하는 로봇이 등장한다. 더 나아가 자율적으로 이동 가능한 자율 이동 로봇도 빠르게 나오고 있다. 이 로봇은 계획을 수립하고 행동하고, 불완전한 환경에는 스스로 대처가 가능한 로봇이다. 미래의 로봇은 인간형 로봇, 휴머노이드를 생각할 수 있는데, 인간처럼 손, 발, 눈, 손가락을 가지고 주위 환경에 인간처럼 반응하는 로봇, 우리가 스타워즈를 보면, 감탄을 했던 그런 로봇들인 나올 것이다. 정말 영화에서처럼 클론과의 전쟁을 할지도 모르고, 내가 기계 인간이 될지도 모른다.

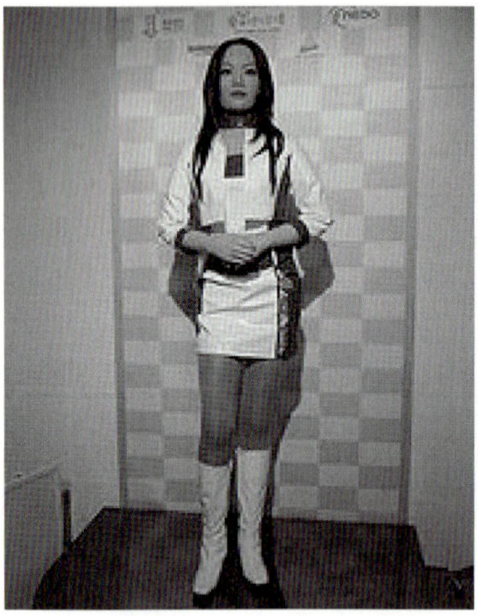

유비쿼터스 로봇은 독립형 로봇과 네트워크 로봇으로 나누어 볼 수 있다.
독립형 로봇은 네트워크에 연결되어 있지 않고, 로봇 자체적으로 센싱하고 처리한 후 행동하는 로봇이다. 구성도를 보면, 다음 그림과 같다.

영어 작문 연습

- Ubiquitous robot can be divided independent robot and network robot. Independent robot is not connected to network and sense, process and act by itself. The constitution is as follows.

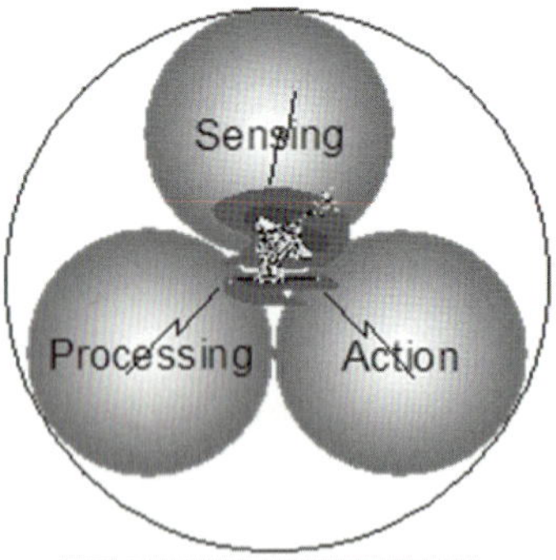

[그림 15] 독립형 로봇

네트워크 로봇은 유비쿼터스 센서 네트워크를 이용해서 센싱된 데이터를 수집을 한다. 처리도 직접하기 않고, 고성능 분산 서버를 이용해서 처리 한 후 그 지시사항을 로봇에 지시를 하면, 그 지시에 따를 뿐이다. 당연히 네트워크에 유무선으로 연동되어 사용자를 돕게 된다.

영어 작문 연습

- Network robot collects data sensed by using Ubiquitous sense network. It does not process by itself but follow the direction of processed result by high effectiveness dispersion server. Naturally it helps users with being interlocked with network wired or wirelessly.

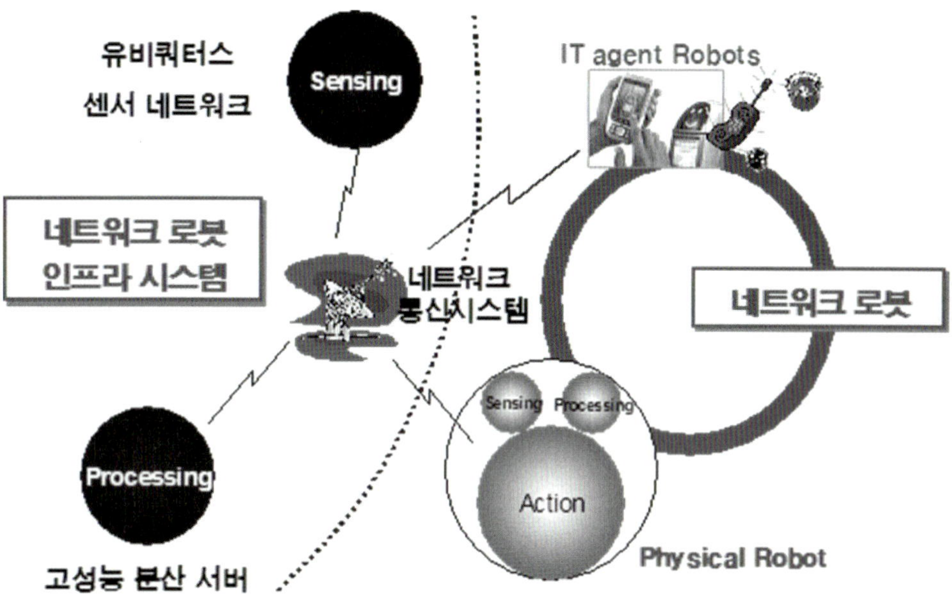

[그림 16] 네트워크 로봇

9.5 Glossary

- Anthrobotics
- Artificial intelligence
- Autonomous research robotics
- BEAM robotics
- Behavior-based robotics
- Biomorphic robotics
- Bionics
- Biorobotics
- Cognitive robotics
- Computer vision
- Machine vision
- Developmental robotics
- Driverless cars
- Electronic Stability Control
- Epigenetic robotics
- Evolutionary robotics
- Home automation
- Human robot interaction
- Intelligent vehicle technologies
- Laboratory robotics
- Nanorobotics
- Passive dynamics
- Programming by Demonstration
- Rapid prototyping
- Robotic surgery
- Robot kinematics
- Robot locomotion
- Speech processing
- Swarm robotics
- Telepresence

References [참고문헌]

[1] 융합 그리고 넥스트, 손병희 저, 교우사, 2015

[2] "Telecom glossary "bot"". Alliance for Telecommunications Solutions. 2001-02-28.

[3] Archived from the original on 2008-07-14.
http://web.archive.org/web/20070202121608/http://www.atis.org/tg2k/_bot.html. http://www.emrotechnologies.com
Ho, C. C.; MacDorman, K. F.; Pramono, Z. A. D. (2008). "Human emotion and the uncanny valley: A GLM, MDS, and ISOMAP analysis of robot video ratings". Proceedings of the Third ACM/IEEE International Conference on Human-Robot Interaction. March 11-14. Amsterdam.

[4] Cambridge Advanced Learner's Dictionary, Cambridge University Press

[5] Collins COBUILD Intermediate Learner's Dictionary, Heinle &Heinle Publishers

[6] Longman Essential Activator, Longman National Corpus

[7] English Grammar in Use, Raymond Murphy, Cambridge University Press

9과 연습문제

01 Match each word in the first column with its definition in the second column.

1.	consensus	A.	imitate
2.	mimic	B.	independent
3.	autonomous	C.	general agreement
4.	Assembly	D.	the process of putting together the parts
5.	depiction	E.	kind and gentle
6.	benign	F.	the way that something is shown

02 Each of the following excerpts is about robots. Read the sentences. Fill in the blanks using the words in the below box.

1) A ________________ is a virtual or mechanical artificial agent. In practice, it is usually an electro-mechanical ________________ which is guided by computer or electronic programming, and is thus able to do tasks on its own. Another common characteristic is that by its appearance or ________________, a robot often conveys a sense that it has intent or agency of its own.

movements, robot, machine

2) Typical ________________ of robots include welding, painting, assembly, pick and place, packaging and palletizing, product inspection, and testing, all accomplished with high ________________, speed, and ________________.

endurance, applications, precision

3) ________________ have captured the imagination of inventors and storytellers for centuries. While the birth of automation ________________ can be traced to the Greeks, the first industrial robots were not produced until the 1970's. Today's cutting edge ________________ robots are the result of years of trial and error.

robots, humanoid, technology

4) Some ________________ are programmed to faithfully carry out specific actions over and over again (repetitive actions) without variation and with a high degree of ________________. These actions are determined by programmed routines that specify the direction, acceleration, velocity, deceleration, and distance of a series of coordinated ________________.

accuracy, motions, robots

5) Perhaps the most popular ________________ of robots is in industrial welding. The repeatability, uniformity quality, and speed of robotic welding is unmatched. The two basic ________________ of welding are spot welding and arc welding, although laser welding is done. Some environmental requirements should be considered for a successful ________________.

types, applications, operation

6) Who ________________ the term robots? In 1956, George Devil and Joseph Engelberger formed the world's first robot ________________, but writers have been dreaming about ________________ long before that.

company, robots, invented

7) An industrial ________________ is officially defined by ISO as an automatically controlled, reprogrammable, multipurpose manipulator programmable in three or more axes. The field of ________________ may be more practically defined as the study, design and use of robot systems for manufacturing (a top-level definition relying on the prior ________________ of robot).

robot, definition, robotics

8) Modern industrial robots are true marvels of ________________. A robot the size of a person can easily carry a load over one hundred pounds and move it very quickly with a repeatability of +/−0.006 inches. Furthermore these ________________ can do that 24 hours a day for years on end with no failures whatsoever. Though they are reprogrammable, in many ________________ (particularly those in the auto industry) they are programmed once and then repeat that exact same task for years.

engineering, robots, applications

9) ________________ are used in a wide range of industrial applications. The earliest ________________ were in materials handling, spot welding, and spray painting. Robots were initially ________________ to jobs that were hot, heavy, and hazardous such as die casting, forging, and spot welding.

applications, robots, applied

10) Interest in ________________ increased in the late 1970s and many US companies entered the field, including large firms like General Electric, and General Motors (which formed joint venture FANUC Robotics with FANUC LTD of Japan). U.S. startup companies included Automatix and Adept Technology, Inc. At the height of the robot boom in 1984, Unimation was acquired by Westinghouse Electric

Corporation for 107 million U.S. dollars. Westinghouse sold Unimation to Stäubli Faverges SCA of France in 1988, which is still making ________________ robots for general industrial and cleanroom applications and even bought the robotic ________________ of Bosch in late 2004.

articulated, division, robotics

10

Digital TV

10 Digital TV

10.1 Introduction

Digital television (DTV) is the sending and receiving of moving images and sound by **discrete** (digital) signals, in contrast to the analog signals used by analog TV.

The first country to make a wholesale switch to digital over-the-air (terrestrial) broadcasting was Luxembourg, in 2006. Since then, the Netherlands, Finland, Andorra, Sweden, Switzerland, Germany and Portugal have **followed suit**.

In the United States, where most full-power television stations are already broadcasting both an analog and a digital signal, the FCC has set June 12, 2009 as the date on which all analog transmissions will cease and over-the-air broadcasts will be solely digital.[1] By special **dispensation**, some analog TV signals ceased, as previously scheduled, on February 17, 2009.[2] This **has to do with** station resistance to a sudden change in schedule, that will cost broadcasters money and cause logistical nightmares.

In Japan, the **switch** to digital is scheduled to happen July 24, 2011. In Canada, it is scheduled to happen August 31, 2011. China is scheduled to switch in 2015. In the United Kingdom, the digital switchover has different times for each part of the country; however, the whole of the UK will be digital by 2012. Brazil switched to digital on December 2, 2007 in major cities and it is estimated it will take seven years for complete signal expansion over all of the Brazilian territory.

In Malaysia, the Malaysian Communications & Multimedia Commission (MCMC) will **call for** tender bids in the third quarter of 2009 for the UHF 470–742 megahertz spectrum which will **pave the way for** the country to move into the digital television era. The awarding of the spectrum will see the winner having to build a single digital terrestrial transmission/TV broadcast (DTTB) infrastructure for all broadcasters to ride on to transmit their TV programs. The winner will be announced at the end of 2009 or early 2010 and has to **commence** digital **roll-out** soon after the award where the analog switch-off is planned for 2015.

While the majority of the viewers of over-the-air broadcasting in the USA watch full-power stations (which number about 1800), there are three other categories of TV stations in the USA : low-power stations, Class A stations, and TV translator stations. There is presently no deadline for these stations, about 7100 in number, to **convert to** digital broadcasting.

- From Wikipedia, the free encyclopedia

이번 장에서 꼭 알아야 할 단어

discrete

분리된, 별개의

= Having a clear independent shape or form; separate

- These small companies now have their own **discrete** identity.
- The universe is composed of **discrete** bodies.
- Computers treat time as a series of **discrete** moments rather than a continuous flow.

switch

변화

*A **switch** is a sudden or complete change, especially of one person or thing for another.*

- She started studying English at college but **switched** to Business Studies in her second year.
- In 1971, Britain **switched** over to a decimal currency.
- After the bank robbery, the gang **switched** cars.

over-the-air

방송으로

***Over-The-Air** or **OTA** refers to signals sent by a broadcast television station through the open air.*

☞ *Over The Air (OTA) (or Over-The-Air) is a standard for the transmission and reception of application-related information in a wireless communications system.*

☞ *Over-the-air programming(OTA) may refer to either free-to-air, terrestrial television, or in the mobile content world, over-the-air service provisioning(OTASP), over-the-air provisioning(OTAP) or over-the-air parameter administration(OTAPA), methods of distributing new software updates to cellphones or provisioning handsets with the necessary settings with which to access services such as WAP or MMS.*

follow suit

(다른 사람이 하는 대로) 따라하다

*If people **follow suit**, they do what someone else has just done.*

- I tent to avoid such companies, and I hope other people will **follow suit.**
- Other manufacturers are beginning to **follow suit** and release green power supplies, so check the market often to see what's available.
- From that point on, the personal computer really began to take off. Other manufacturers **followed suit**, and soon there were computers from Commodore, Atari and Texas Instruments.

dispensation

특별 허가

*A **dispensation** is special permission to do something that is normally not allowed.*

- They were promised **dispensation** from military service.
- The couple has requested special **dispensation** from the Church to allow them to marry.
- The Vatican granted a special **dispensation** for the Mass.

A has to do with B

A 와 B 는 관련 있다

= A is connected or concerned with B

- It **has to do with** giving back.
- The decision **has to do with** my marriage.
- What does economics **have to do with** politics?

call for

~에 대한 요구

*If there is a **call for** something, someone demands that it should happen.*

- a **call for** all businessmen to work.
- This celebration **calls for** balloons.
- This job **calls for** a strong commitment from employees.

pave the way for

~을 위해 길을 열다

*If one thing **paves the way for** another, it creats a situation in which the other thing is able to happen.*

- UCSD researchers **pave the way for** effective liver treatments.
- A peace agreement last year **paved the way for** this week's elections.
- Scientists hope that data from the probe will **pave the way for** a more detailed exploration of Mars.

commence

~이 시작되다(하다)

*When something **commences,** or when you commence it, it begins.*

- Shall we let the meeting **commence,** gentlemen?
- We will **commence** building work in August of next year.
- Unfortunately, he **commenced** speaking before all the guests had finished eating.

roll-out

첫 공개, 출시

Making a new product, service, or system available for the first time

- The **roll-out** is due to be completed by 17 November.
- However, media observers say that testing the waters like this before a major global **roll-out** is typical of its strategy.

convert (in)to

~로의 전환(변화)

*To **convert** one thing **into** another means to change it into a different shape or form.*

- Could we **convert** the small bedroom **into** a second bathroom?
- What's the formula for **converting** pounds **into** kilos?
- He **converted to** Catholicism when he got married.

Common Collocations

Discrete	Switch	over-the-air
Discrete math Discrete variable Discrete data	Switch case Switch anime Switch java	Over-the-air channel listing Over-the-air HD antenna Over-the-air update
Dispensation	**call for**	**convert to**
Papal dispensation Dispensation theology Dispensation marriage	Call for submissions Call for papers Call for artists	Convert to Excel Convert to hwp Convert to doc

10.2 Useful Expressions

in contrast to : ~와는 다르게, ~와는 대조적으로	
• In contrast to their neighbors, they live modestly. • In constrast to her coworkers, she never stays late. • In contrast to his smart older brother, he's dumb.	• 그들의 이웃과는 다르게, 그들은 겸손하게 산다. • 그녀의 동료들과는 다르게, 그녀는 절대 늦게까지 남지 않는다. • 그의 똑똑한 형과는 대조적으로, 그는 바보 같다.
deadline for : ~에 대한 마감시한	
• There is no deadline for this project. • What is the deadline for applying for this position? • The deadline for applications is 30 April.	• 이 프로젝트는 마감이 없다. • 이 직책에 지원하기 위한 마감이 언제지? • 지원마감일자는 4월 30일입니다.

10.3 Descriptive Grammar

In the United States, *where* most full-power television stations are already broadcasting both an analog and a digital signal, ...

관계부사 where은 '~에서(in which)'라는 의미를 가지고 있다. 'Which'는 바로 앞에 나온 명사를 대신해서 사용되는 일종의 대명사라고 생각하면 된다. 예를 들어, 'This is the city *where* I was born.' 이라는 예문에서 where은 'in which'인데, 여기서 which는 바로 앞에 나온 명사, 즉 'the city'를 의미한다고 보면 된다. 그래서 이 문장에서 where은 'in the city'라는 의미로 보고 지나가면 되는 것이다. 의미를 해석해 보면, '여기가 도시이다 / *이 도시에서* / 내가 태어났다.' 가 된다.

Where은 장소 뿐만 아니라 상황이나 조건을 수식할 때도 많이 쓰인다. 아래 예문을 통해서 where의 쓰임에 대해 더 알아보도록 하자.

- This is the place *where* we used to play.
- This is a situation *where* a lot of problems could occur.
- When a person is in a situation *where* they are unsure of the correct way to behave, they will often look to others for cues concerning the correct behavior.

위의 문장을 다시 살펴보면,

In the United States, / where most full-power television stations are already broadcasting both an analog and a digital signal, . . .
미국에서는, / 대부분의 전출력 TV 방송국들이 이미 아나로그와 디지털 시그널로 방송하고 있는. . .

10.4 Substance

World map of digital television transition progress.

Legend

- Transition completed, all analog signals terminated
- Transition in progress, broadcasting both analog and digital signals
- Transition not yet started, broadcasting analog signals only
- Does not intend to transition, broadcasting analog signals only
- No information available

방송 컨텐츠의 제작, 전송 및 재현에 이르는 모든 과정을 디지털화 함으로써, 고품질의 방송 프로그램 및 다채널 서비스를 실현하고 방송통신 융합형 서비스를 포함한 다양한 부가기능을 제공하는 방송 기술이다. 디지털 TV/방송의 개념도는 다음과 같다.

영어 작문 연습

- Digital TV/broadcast is the technology of accomplishing high quality broadcast program and service and supplying various additional function including service integrated with communication by digitalizing all processes such as production, sending, and, reproduction of contents. The concept figure of digital TV/broadcast is as follows.

[그림 17] 디지털TV/방송의 개념도

이제 아날로그 TV에도 변환 장치 (셋톱 박스)와 안테나만 달면, 디지털 TV를 볼 수 있다. 현재 아날로그TV 보다 5 ~ 6배 선명한 고선명 (HD) 영상과 CD급 고음질의 음향이 제공된다. 이렇게 되므로써, 언제 어디서나 실감나는 TV, 대화형 TV로 TV가 똑똑해 지고 있다. 거기에, 시간, 공간적 제약없이 영상, 음성, 음향, 문자, 그래픽 등 다양한 형태의 정보와 TV 수상기, 컴퓨터 모니터, 이동통신단말 등 여러 가지 매체를 통합하여 멀티미디어 방송과 시청자가 참여하는 양방향 서비스 기능이 가능하다.

고속 이동 중에도 선명한 방송시청이 가능한 DMB방송은 기존 방송의 공간적 한계를 극복하며, 통신망과 연계된 융합형 차세대 멀티미디어 서비스를 제공한다. 하지만, 유료화 논란으로 시장이 복잡하게 얽혀있긴 하다.

디지털 TV는 양방향 데이터 방송, TV전자상거래, T-Government 등 멀티미디어 부가서비스를 이용하여 디지털 인프라간 통합의 구심점 및 정보 플랫폼 (Home Gateway) 역할을 수행할 것이다.

영어 작문 연습

- Now we can watch digital TV if only we install inverter (set top box) and antenna on analogue TV. At present, 5–6 times clearer high density (HD) image and high quality sound like CD is offered. With this change, TV is getting smarter which is true to nature anytime and anywhere and interactive.
- And what's more, 2-way service multimedia broadcast and TV audience participated in is possible without limit of time and space with various forms of information like image, voice, sound, letter, graphic, and etc. and unified medias like TV set, computer monitor, mobile terminal, and etc.
- Among high speed mobiles DMB broadcast offering clear image overcome spatial boundary of existing broadcast and supply next generation multimedia service integrated with communication network. But it is entangled with dispute about charge complicatedly in market.
- Digital TV will carry out the glue of unification among digital infra and information platform (Home Gateway) using multimedia additional service such as 2-way data broadcast, TV electronic commerce, T-Government and etc.

10.5 Glossary

- ATSC Standards
- ATSC tuner
- Broadcast television systems
- Digital radio, section List of digital television broadcasting systems
- OFDM system comparison table
- Spectral efficiency comparison table including digital television broadcasting
- Digital television transition
- Digital terrestrial television in Australia
- Digital terrestrial television in Ireland
- Digital terrestrial television in the United Kingdom
- DMB-T/H, China's digital television standard
- DVB-T, Digital Video Broadcasting – Terrestrial
- DTV transition in the United States
- Electronic Programme Guide
- Gigaset
- High-definition television

- Interactive television
- ISDB, Japan's digital television standard
- LinuxTV
- List of digital television deployments by country
- Multimedia Home Platform (MHP) and Zapper
- North American broadcast television frequencies
- Personal video recorder
- Redesign project
- SBTVD, Brazil's digital television standard
- Set-top box (STB)
- System-on-a-chip
- TS file format

References [참고문헌]

[1] 융합 그리고 넥스트, 손병희 저, 교우사, 2015

[2] Cambridge Advanced Learner's Dictionary, Cambridge University Press

[3] Collins COBUILD Intermediate Learner's Dictionary, Heinle & Heinle Publishers

[4] Longman Essential Activator, Longman National Corpus

[5] English Grammar in Use, Raymond Murphy, Cambridge University Press

10과 연습문제

01 Match each word in the first column with its definition in the second column.

1. discrete	A. change
2. switch	B. relate to
3. dispensation	C. begin
4. has to do with	D. separate
5. call for	E. a special permission
6. commence	F. demand

02 Each of the following excerpts is about digital television (DTV). Read the sentences. Fill in the blanks using the words in the below box.

1) Digital ____________ (DTV) is the sending and receiving of moving images and sound by discrete (digital) ____________, in contrast to the ____________ signals used by analog TV.

signals, television, analog

2) Your DTV ____________ can be affected by terrain, trees, buildings, the weather, damaged equipment, as well as antenna type, location, and orientation. It can be improved just by changing the location of the ____________ you're using now. Moving your antenna away from other objects and structures, or placing it higher, can often improve reception. The performance of outdoor antennas can degrade over time due to exposure to the weather. Also, you may consider installing a signal booster, which should ____________

reception.

improve, reception, antenna

3) Digital ________________ supports many different picture formats defined by the combination of size, aspect ratio (height to width ratio) and interlacing. With terrestrial ________________ in the USA, the range of formats can be coarsely ________________ into two categories : HDTV and SDTV.

divided, television, broadcasting

4) ________________ television (DTV) is an innovative ________________ of over-the-air broadcasting technology that enables TV stations to provide dramatically clearer ________________ and better sound quality.

type, pictures, digital

5) ________________ commenced in 2001 in metropolitan areas and digital signals have been broadcast in many regional areas since then. Digital ________________ are currently transmitted alongside analog broadcasts, but when the analog signals are switched off, only TVs which are capable of receiving digital ________________ through a set top box or built-in tuner will work.

signals, digital TV, broadcasts

6) Digital ________________, or DTV, is a ________________that gives viewers crystal-clear pictures and sound, and more ________________choices than ever before.

technology, programming, television

7) Your DTV reception can be affected by nearby moving vehicles, such as cars, trucks, trains and airplanes. In some instances, shadowing or reflections from these vehicles may cause your digital ________________ to temporarily break-up or even disappear completely. If this occurs, you should try moving or reorienting your ________________ to find a position that provides the most reliable reception. If you are using an indoor antenna, switching to an outdoor antenna system which may include a directional antenna or rotor could improve reception. In severe cases it may not be possible to completely eliminate the effect of nearby traffic. If ________________ remains unsatisfactory due to these disruptions, viewers may wish to consider alternatives such as cable or satellite service.

antenna, picture, reception

8) High-definition television (________________), one of several different formats that can be transmitted over DTV, uses one of two formats: 1280 × 720 pixels in progressive scan mode (abbreviated 720p) or 1920 × 1080 ________________ in interlace mode (1080i). Each of these utilizes a 16:9 aspect ratio. (Some televisions are capable of receiving an HD resolution of 1920 × 1080 at a 60 Hz progressive scan frame rate — known as 1080p60, but this standard is not currently used for transmission.) HDTV cannot be transmitted over current ________________ channels.

analog, pixels, HDTV

9) DTV is more efficient and more flexible than the ________________ way of broadcasting known as analog. For example, ________________ makes it possible for stations to broadcast multiple channels of free programming all at once (called multicasting), instead of broadcasting one channel at a time. DTV ________________ can also be used to deliver future interactive

video and data services that analog technology can't provide.

technology, traditional, DTV

10) There are a number of ________________ ways to receive digital television. One of the oldest means of receiving DTV (and TV in general) is using an antenna (known as an aerial in some countries). This way is known as Digital Terrestrial ________________ (DTT). With DTT, viewers are limited to whatever channels the ____________________ picks up. Signal quality will also vary.

different, television, antenna

11

Content Management

11 Content Management

11.1 Introduction

Content management, or CM, is a set of processes and technologies that support the **evolutionary** life cycle of digital information. This digital information is often referred to as content or, to be precise, digital content. Digital content may **take the form of** text, such as documents, multimedia files, such as audio or video files, or any other file type which follows a content lifecycle which requires management.

- From Wikipedia, the free encyclopedia

이번 장에서 꼭 알아야 할 단어

evolutionary

점진적인

Evolutionary *means relating to a process of gradual change and development.*

- The change has been **evolutionary** rather than revolutionary.
- We believe that our proposals are practical, **evolutionary** and based on markets and choice.

take the form of

~한 모습(형태)을 갖게 되다

= to gradually be seen or gradually develop.

- Concepts may **take the form of** ideas, mockups or prototypes.
- As they cheated, the idea of a holiday together gradually **took form.**
- Viral promotions may **take the form of** video clips, interactive Flash games, advergames, ebooks, brandable software, images, or even text messages.

Common Collocations

Evolutionary
Evolutionary biology Evolutionary algorithm Evolutionary theory Evolutionary fitness

11.2 Useful Expressions

a set of : 일련의	
• The Foreign Minister will announce a set of measures tomorrow. • An encyclopedia has been published as a set of books. • Candence is a set of rhythm in a dance or a march.	• 외교부 장관은 내일 종합 대책을 발표할 것이다. • 백과사전은 여러 권의 책으로 출판되어 왔다. • 칸덴스는 댄스곡이나 행진곡에 나오는 일련의 리듬이다.
To be precise : 더 정확히(엄밀히) 말하면	
• To be precise, the negotiations are being prepared • To be precise, I was not a member of the Government; nor was I a Member of Parliament at the time. • I have a large number of mobile homes in my area--614 to be precise.	• 더 정확히 말하자면, 협상은 진행 중에 있다. • 정확히 말하자면, 나는 당시에 정부의 일원도 아니었고, 의회에 소속되어 있지도 않았었다. • 나는 내 주변에 정말 많은 이동주택들이 있다 – 정확히 614개의 이동주택들이 있다.

11.3 Descriptive Grammar

Content management, or CM, is a set of processes and technologies *that* support the evolutionary life cycle of digital information.

관계대명사[who/which/that]는 절[주어와 동사를 가지고 있는 문장]의 형태로 명사를 수식해 주는 역할을 한다는 것을 9강에서 다뤘었다. 이번 장에서는 관계대명사가 문장 내에서 주어 역할을 하는 경우를 살펴보도록 하자.

- I thank the woman *who(that)* helped me.

예문을 보면, 'I thank the woman.'과 'The woman helped me.' 두 문장이 결합된 것이라고 볼 수 있다. 즉 두 문장을 연결해 주는 관계대명사 who는 문장에서 The woman이라는 주어를 대신한다고 볼 수 있다.

- The book is mine.
- The book is on the table.

이 두 문장을 관계대명사를 사용하여 한 문장으로 만든다면 다음과 같이 만들 수 있을 것이다.

The book *which(that)* is on the table is mine.
참고로 관계대명사 that은 선행사가 사람, 사물인 경우 모두 사용 가능하다.

위의 문장을 다시 살펴보면,

Content management, or CM, is a set of processes and technologies / *that* support / the evolutionary life cycle of digital information.
콘텐츠관리, 혹은 CM이라고도 하는 이것은 / 일련의 과정 그리고 기술이다 / (일련의 과정과 기술이) 지원해 준다 / 디지털 정보의 점차적인 라이프 주기를

11.4 Substance

디지털 컨텐츠는 부호·문자·음성·음향·이미지 또는 영상 등으로 표현된 자료 또는 정보로서 그 보존 및 이용에 있어서 효용을 높일 수 있도록 전자적 형태로 제작 또는 처리된 자료다. (온라인디지털컨텐츠산업발전법 제2조)

그럼 디지털 컨텐츠 기술이란 무엇인가?

디지털 컨텐츠 기술이란 영상, 음향, 문자 등 다양한 형태의 정보 컨텐츠를 디지털화 된 서비스나 제품의 형태로 가공, 생산, 유통하기 위한 기술이다.

영어 작문 연습

- Digital contents is data or information expressed to code, letter, voice, sound, image and reflection and produced and processed in the electronic form to enhance the effectiveness in using. (online digital contents industry development the second)
- Then what is the digital contents technology?
- Digital contents technology is the technology for processing, producing and distributing contents of the various form of image, sound, letter, and, etc. to the form of digitalized service or product

개념도는 다음과 같다.

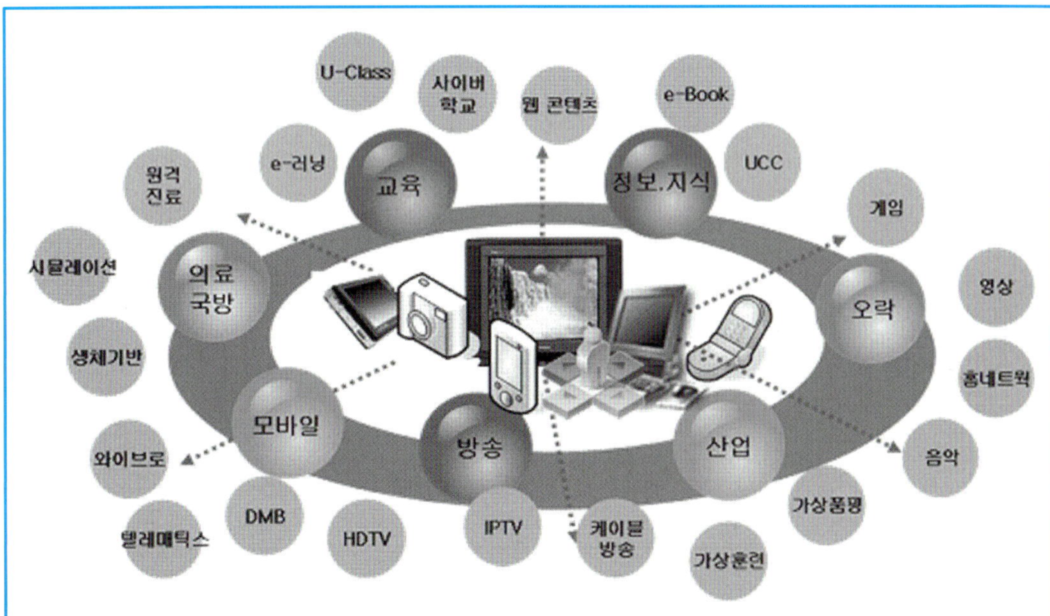

[그림 18] 디지털 컨텐츠 개념도

디지털 컨텐츠 기술은 고도화되는 서비스 환경에 상응하여, 기술의 형태도 보다 복합적이고 다양하게 변화되어 가고 있다.

영어 작문 연습

- Digital contents technology is changing multiply and variously in technical form corresponding to the highly developing service environment.

11.5 Structure

Positive

Conjugate the helping verb "be" + verb + -ing.
I'm (I am) → working today.
You're (You are) → working today.
He's (He is) → working today.
She's (She is) → working today.
It's (It is) → working today.
We're (We are) → working today.
You're (You are) → working today.
They're (They are) → working today.

Negative

Conjugate the helping verb "be" + not + verb + -ing.
I'm not (I am not) → coming this evening.
You aren't (You are not) → coming this evening.
He isn't (He is not) → coming this evening.
She isn't (She is not) → coming this evening.
It isn't (It is not) → coming this evening.
We aren't (We are not) → coming this evening.
You aren't (You are not) → coming this evening.
They aren't (They are not) → coming this evening.

Questions

Question word + conjugate the helping verb 'be' + subject + verb + -ing

What → are you → doing this afternoon?

What → is he → doing this afternoon?

What → is she → doing this afternoon?

What → is it → doing this afternoon?

What → are we → doing this afternoon?

What → are you → doing this afternoon?

What → are they → doing this afternoon?

11.6 Glossary

- List of content management systems
- Enterprise content management
- Enterprise information management
- Content management system (includes links to system providers)
- Information architecture
- Website architecture
- Web design
- Digital asset management
- Content engineering
- Content delivery
- Content Management Interoperability Services

References [참고문헌]

[1] 융합 그리고 넥스트, 손병희 저, 교우사, 2015

[2] Cambridge Advanced Learner's Dictionary, Cambridge University Press

[3] Collins COBUILD Intermediate Learner's Dictionary, Heinle & Heinle Publishers

[4] Longman Essential Activator, Longman National Corpus

[5] English Grammar in Use, Raymond Murphy, Cambridge University Press

11과 연습문제

01 Match each word in the first column with its definition in the second column.

1.	evolutionary	A.	gradually develop.
2.	to be precise	B.	relating to a process of gradual change or development
3.	take the form of	C.	to be specific

02 Each of the following excerpts is about content management. Read the sentences. Fill in the blanks using the words in the below box.

1) A content management system (_______________) such as a document management _______________ (DMS) is a computer application used to manage work flow needed to collaboratively create, edit, review, index, search, publish and archive various kinds of _______________ media and electronic text.

digital, CMS, system

2) Content Management _______________ come in all shapes and sizes. Chances are, if you've put up a Web _______________ on a free Web page server or ISP you've had access to a rudimentary content _______________ system.

management, systems, page

3) Oracle Universal Content Management (UCM) is the industry's most unified enterprise content management _______________ that enables you to leverage market-leading document management, Web content

management, digital asset management, and records retention functionality to build and complement your business applications. Building a strategic enterprise ________________ management infrastructure for content and applications helps you to reduce costs, easily share content across the enterprise, minimize risk, automate expensive, time-intensive and manual processes, and consolidate multiple Web sites onto a single platform for centralized management. Through user-friendly interfaces, roles-based authentication and security models, Oracle Universal Content ________________ empowers users throughout the enterprise to view, collaborate on or retire content, ensuring that all accessible distributed or published information is secure, accurate and up-to-date.

platform, content, management

4) There are two fundamental ________________ to content management: (1) storing stuff in a content ________________, and (2) supporting the workflow of a group of people ________________ in putting stuff into that repository.

repository, elements, engaged

5) CMSs are frequently used for storing, controlling, versioning, and publishing industry-specific ________________ such as news articles, operators' manuals, technical manuals, sales guides, and marketing brochures. The ________________ managed may include computer files, image media, audio files, video files, electronic documents, and Web content. These concepts represent integrated and interdependent layers. There are various nomenclatures known in this area: Web Content Management, Digital Asset Management, Digital Records Management, Electronic Content Management and so on. The bottom line for these ________________ is managing content and publishing, with a workflow if required.

content, systems, documentation

6) Effective content ________________ systems support an organization's business processes for acquiring, filtering, organizing, and controlling access to information. Because no two ________________ use identical business processes, content management ________________ in practice can be compared to snowflakes -- no two such systems will look or behave exactly alike!

systems, organizations, management

7) YouTube has created an advanced set of copyright policies and ________________management tools to give rights holders ________________ of their content. YouTube provides content management ________________for rights holders of all sizes across the world, and provides tools to cater to the specific needs of various rights owners.

content, solutions, control

8) ________________, or CM, is the set of processes and technologies that support the collection, managing, and publishing of information in any form or medium. In recent times this information is typically referred to as content or, to be precise, ________________ content. Digital content may take the form of text, such as documents, multimedia files, such as audio or video ________________, or any other file type which follows a content lifecycle which requires management.

files, content management, digital

9) There are plenty of ________________when it comes to picking a content management system for a development ________________. Depending on how advanced you need the CMS to be, what language it's built in, and who is going to be using it, it can be a nightmare trying to find the "perfect" ________________for a project.

CMS, options, project

10) Content ______________ practices and goals vary with mission. News organizations, e-commerce websites, and educational institutions all use ______________ management, but in different ways. This leads to differences in terminology and in the names and number of steps in the ______________.

management, content, process

12

Security

12 Security

12.1 Introduction

Security is the degree of protection against danger, loss, and criminals. Individuals or actions that **encroach** upon the condition of protection are responsible for a "**breach** of security."

A condition that results from the establishment and maintenance of protective measures that ensures a state of **inviolability** from hostile acts or influences.

Security has to be compared and contrasted with other related concepts: Safety, continuity, reliability. The key difference between security and reliability is that security must take into account the actions of people attempting to cause destruction.

Computer security is a branch of technology known as information security as applied to computers. The objective of computer security includes protection of information and property from theft, **corruption**, or natural disaster, while allowing the information and property to remain accessible and productive to its intended users.

- From Wikipedia, the free encyclopedia

이번 장에서 꼭 알아야 할 단어

encroach

~을 침해하다, 잠식하다

*If one thing **encroaches** on another, the first thing spreads or becomes stronger, and slowly begins to restrict the power, range, or effectiveness of the second thing.*

- The new institutions do not **encroach** on political power.
- The movie industry had chosen to ignore the **encroaching** competition of television.
- Throughout human history there are numerous examples of one people **encroaching** on the territory of others.

breach

위반

*A **breach** of an agreement, a law, or a promise is an act of breaking it.*

- They felt that our discussions with other companies constituted a **breach** of our agreement.
- He was sued for **breach** of contract.
- There have been serious security **breaches**.

inviolability

불가침

*If a law or principle is **inviolable**, you must not break it.*
☞ ***invilolability**는 명사형임*

- One basic principle would be the **inviolability** of the national borders.
- This essay explores the issues related to the **inviolability** of the human body.

corruption

부패, 타락

__Corruption__ is dishonesty and illegal behavior by people in positions of authority or power.

- **Data corruption** refers to errors in computer data that occur during transmission, retrieval, or processing, introducing unintended changes to the original data.
- Investors demand high-tech solution to end **corruption.**
- File **corruption** occurs whenever data becomes damaged in some way.

Common Collocations

Breach	Corruption
Breach of trust Breach of fiduciary duty Breach of contract	Business corruption Corruption index Police corruption Political corruption

12.2 Useful Expressions

take into account[consideration] : ~을 고려하다, ~을 참작하다	
• I will properly take into account the reserve powers. • Would you take my suggestion into account? • There is an attempt to take all the evidence into account. • He did not take my age into consideration [account].	• 나는 적절히 여력을 참작할 것이다. • 제 제안을 고려해 주시겠습니까? • 이 모든 증거들을 참작하려고 시도가 있습니다. • 그는 내 나이를 고려하지 않았다.
be[become] known as : ~로 알려져 있다[알려지게 되었다].	
• He is known as the smart one around here. • Soybeans are well known as very nutritious food. • Lake Baikal is known as the deepest lake on the earth. • For her compassion, she has become known as an angel of a girl.	• 그는 이 주변에서 똑똑한 사람으로 알려져 있다. • 콩은 매우 영양가 있는 음식으로 알려져 있다. • 바이칼 호수는 지구상에서 가장 깊은 호수로 알려져 있다. • 그녀의 동정심 때문에 그녀는 천사 소녀로 알려져 있다.

12.3 Descriptive Grammar

The objective of computer security includes protection of information and property from theft, corruption, or natural disaster, while allowing the information and property to *remain accessible* and productive to its intended users.

"He is married"라는 문장처럼 주어 + 동사 + 주격보어의 형태를 갖는 문장에서 사용되는 동사는 be동사다. 하지만 이러한 문형에서 사용될 수 있는 동사의 종류는 아래와 같이 다양하다.

- lie / hold / stand ： ~일 상태로 있다.
- keep / remain / stay + 형용사 ： ~일 상태로 유지하다.
- get / grow / become / go / come / turn ： ~인 상태가 되다
- seem / appear ： ~인 것 같다.

이 외에도 look smell taste sound feel과 같은 감각동사

예문을 통해 위 동사들의 쓰임을 살펴보도록 하자.

- *Hold* still! Say cheese~[사진찍는 상황에서 "Hold still!"은 움직이지 말고 가만히 있으라는 의미임]
- Please *remain* calm.
- I *got* very used to eating alone soon.
- He *seems* very disappointed.
- Are you making pizza right now? It *smells* great!

위의 문장을 다시 살펴보면,

The objective of computer security / includes protection of information and property / from theft, corruption, or natural disaster, / while allowing the information and property to remain accessible and productive to its intended users.
컴퓨터 보안의 목적은 / 정보와 재산의 보호를 포함한다. / 절도, 부패, 자연재해로부터의 보호를 / 정보와 재산을 허용하면서 / (정보와 재산이) 접근 가능하고 생산적이도록 유지하는 것을 / 허용된 사용자들에게

12.4 Substance

정보통신망의 마비, 개인정보의 유출, 불건전 정보의 유통 등 정보통신 환경을 저해하는 위협과 부작용에 대응할 수 있도록 정보통신 시스템 및 데이터의 기밀성 (정보 유출 방지), 무결성 (데이터 위조 및 변조 방지)을 유지하고 시스템의 가용성을 보장하는 기술이다. 정보보호기술은 사용자 측면에서 개인정보 및 프라이버시 제공, 서비스 및 디바이스 측면에서 안전한 신뢰 서비스 제공, 인프라 환경 측면에서 끊김 없고 이동성이 지원되는 안전한 인프라를 제공하는 기술로 정의할 수 있다. 그 개념도는 다음과 같다.

영어 작문 연습

- Security is the technique keeping information communication system, confidentiality (prevention of information leak) and, integrity (prevention of data forgery and alteration) and guaranteeing capabilities to cope with threats and side effects injurious to information communication environment like a paralysis of information communication network, information leak, unsound information distribution, and etc.
- Information protection technology can be defined as the technique able to offer individual information and privacy in user's side, safe confidential service in the side of service and device, and continuous and mobile safe infra in the side of infra environment. The figure of concept is as follows.

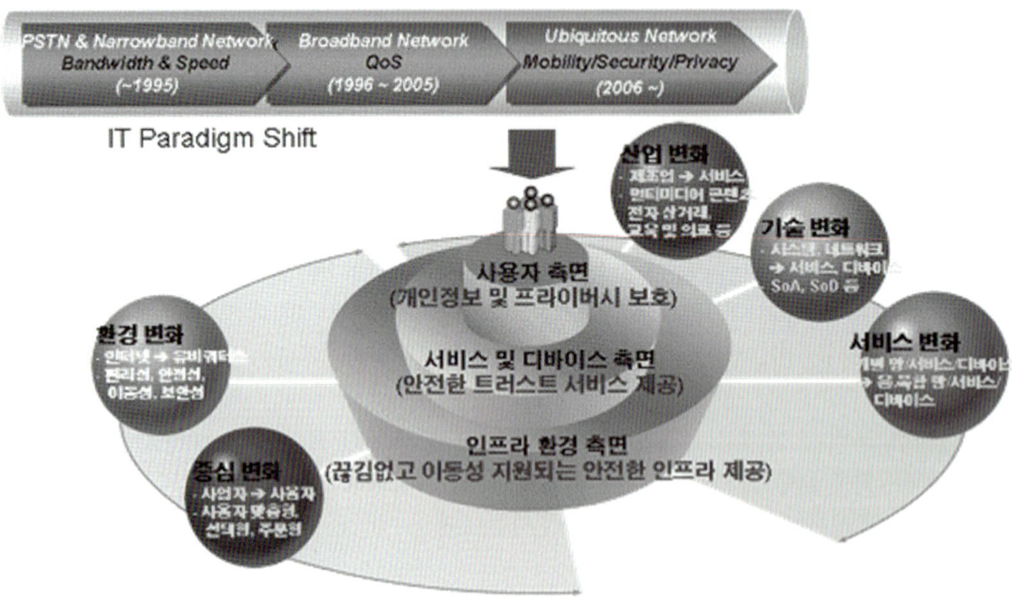

[그림 19] IT 환경 변화에 따른 정보보호

유비쿼터스 정보보호의 목표는 다음과 같다.

- 기밀성 : 정보 누출의 방지한다.
- 무결성 : 정보의 변조 및 파괴를 예방하고 방지한다.
- 가용성 : 해킹으로 인한 시스템 동작 불능을 예방한다.
- 인증 : 정당한 사용자임을 확인한다.
- 책임 추적성 : 행위를 부인하는 것을 부인하는 것을 봉쇄한다.

영어 작문 연습

- The purpose of Ubiquitous information protection is as follows
  - confidentiality : to prevent information leak
  - integrity : to anticipate and prevent information alteration and destruction
  - capability : to anticipate system operation inability by hacking
  - certification : to certify lawful user
  - tracing of responsibility : to block denying one's acting

정보보호 산업 및 기술의 분류는 보는 관점에 따라 다양하게 분류할 수 있다. 우선 다음 그림을 살펴 보자.

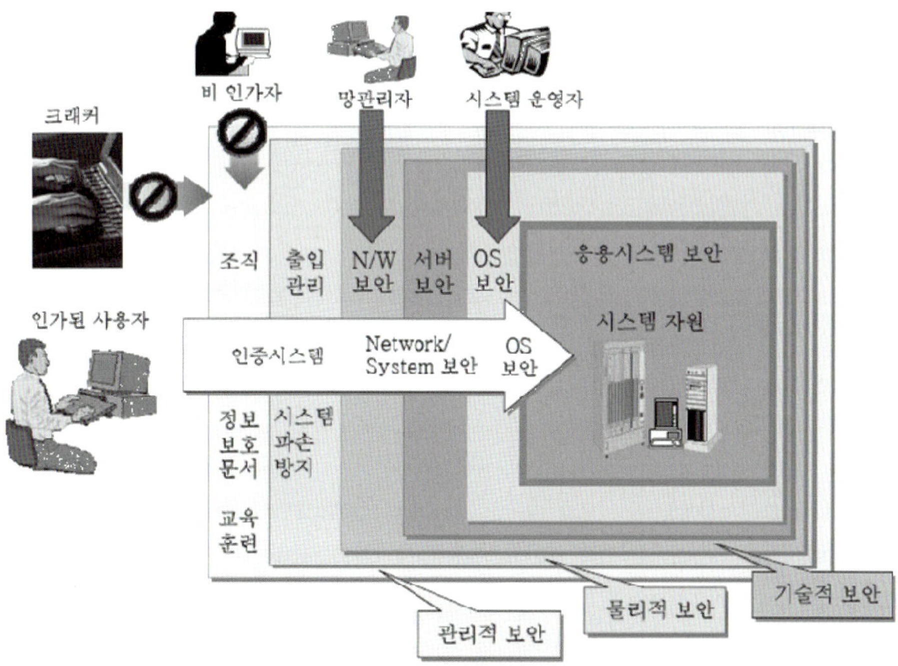

[그림 20] 유비쿼터스 정보보호 분류

먼저, 관리적 보안, 물리적 보안, 기술적 보안이 있다.
관리적 관점에서는 보안 조직, 출입 관리, 교육 훈련, 보안 정책이 있다. 보안 조직에는 시스템 운영자와 망 관리자, 사용자로 나뉘어 보안 등급에 따라 시스템 사용의 차등을 부여하게 되는데, 여기서, 비인가된 사용자는 크래커나 해커로 분류한다.
물리적 관점에서 보면, 시스템 자원 보안, 응용 시스템 자원 보안, OS 보안, 인터넷 보안, 네트워크 보안으로 나뉜다.

영어 작문 연습

- First, we can classify to administrative security, physical security and, technical security.
  In the administrative viewpoint, it can be divided to security configuration, control of entrance and exit, educational training and, security policy. In the security configuration classified to configuration system manager, network administrator, and user receive grade for using the system. Here, a user without permission is classified as a cracker or hacker.
  In the physical viewpoint, it can be divided into system resource security, application system resource security, OS security, internet security and, network security.

12.5 Glossary

The following terms used in engineering secure systems are explained below.

- Authentication techniques can be used to ensure that communication end-points are who they say they are.
- Automated theorem proving and other verification tools can enable critical algorithms and code used in secure systems to be mathematically proven to meet their specifications.
- Capability and access control list techniques can be used to ensure privilege separation and mandatory access control.
- Chain of trust techniques can be used to attempt to ensure that all software loaded has been certified as authentic by the system's designers.
- Cryptographic techniques can be used to defend data in transit between systems, reducing the probability that data exchanged between systems can be intercepted or modified.

- Firewalls can provide some protection from online intrusion.
- Mandatory access control can be used to ensure that privileged access is withdrawn when privileges are revoked. For example, deleting a user account should also stop any processes that are running with that user's privileges.
- Secure cryptoprocessors can be used to leverage physical security techniques into protecting the security of the computer system.
- Microkernels can be reliable against errors: eg EROS and Coyotos.

Some of the following items may belong to the computer insecurity article:

- Access authorization restricts access to a computer to a group of users through the use of authentication systems. These systems can protect either the whole computer – such as through an interactive logon screen – or individual services, such as an FTP server. There are many methods for identifying and authenticating users, such as passwords, identification cards, and, more recently, smart cards and biometric systems.
- Anti-virus software consists of computer programs that attempt to identify, thwart and eliminate computer viruses and other malicious software (malware).
- An application with known security flaws should not be run. Either leave it turned off until it can be patched or otherwise fixed, or delete it and replace it with some other application. Publicly known flaws are the main entry used by worms to automatically break into a system and then spread to other systems connected to it. The security website Secunia provides a search tool for unpatched known flaws in popular products.
- Backups are a way of securing information; they are another copy of all the important computer files kept in another location. These files are kept on hard disks, CD-Rs, CD-RWs, and tapes. Suggested locations for backups are a fireproof, waterproof, and heat proof safe, or in a separate, offsite location than that in which the original files are contained. Some individuals and companies also keep their backups in safe deposit boxes inside bank vaults. There is also a fourth option, which involves using one of the file hosting services that backs up files over the Internet for both business and individuals.
- Encryption is used to protect the message from the eyes of others. It can be done in several ways by switching the characters around, replacing characters

with others, and even removing characters from the message. These have to be used in combination to make the encryption secure enough, that is to say, sufficiently difficult to crack. Public key encryption is a refined and practical way of doing encryption. It allows for example anyone to write a message for a list of recipients, and only those recipients will be able to read that message.

- Firewalls are systems which help protect computers and computer networks from attack and subsequent intrusion by restricting the network traffic which can pass through them, based on a set of system administrator defined rules.
- Honey pots are computers that are either intentionally or unintentionally left vulnerable to attack by crackers. They can be used to catch crackers or fix vulnerabilities.
- Intrusion-detection systems can scan a network for people that are on the network but who should not be there or are doing things that they should not be doing, for example trying a lot of passwords to gain access to the network.
- Pinging The ping application can be used by potential crackers to find if an IP address is reachable. If a cracker finds a computer they can try a port scan to detect and attack services on that computer.
- Social engineering awareness keeps employees aware of the dangers of social engineering and/or having a policy in place to prevent social engineering can reduce successful breaches of the network and servers.
- File Integrity Monitors are tools used to detect changes in the integrity of systems and files.

References [참고문헌]

[1] 융합 그리고 넥스트, 손병희 저, 교우사, 2015
[2] Cambridge Advanced Learner's Dictionary, Cambridge University Press
[3] Collins COBUILD Intermediate Learner's Dictionary, Heinle & Heinle Publishers
[4] Longman Essential Activator, Longman National Corpus
[5] English Grammar in Use, Raymond Murphy, Cambridge University Press

12과 연습문제

01 Match each word in the first column with its definition in the second column.

1.	breach	A.	famous
2.	corruption	B.	an act of breaking an agreement, a law, or a promise
3.	encroach	C.	dishonesty and illegal behavior
4.	take into account	D.	take over
5.	be known	E.	consider

02 Each of the following excerpts is about security. Read the sentences. Fill in the blanks using the words in the below box.

1) Whether your ________________ runs Microsoft® Windows®, Apple's Mac OS, LINUX, or something else, the issues are the same and will remain so as new versions of your ________________ are released. The key is to understand the ________________ -related problems that you need to think about and solve.

computer, security, system

2) Computer ________________ is a branch of technology known as information security as applied to computers and networks. The objective of computer security includes protection of information and property from theft, corruption, or natural disaster, while allowing the ________________ and property to remain accessible and productive to its intended users. The terms computer system security, means the collective processes and mechanisms by which sensitive and valuable

information and services are protected from publication, tampering or collapse by unauthorized activities or untrustworthy individuals and unplanned _______________ respectively.

events, security, information

3) No matter how a home _______________ is connected to the Internet, intruders' attacks are often successful. Many home computer owners don't realize that they need to pay attention to computer _______________. In the same way that you are responsible for having insurance when you drive a car, you need to also be responsible for your home computer's security. This document explains how some parts of the Internet work and then describes tasks you can do to improve the security of your home computer system. The goal is to keep intruders and their _______________off your computer.

programs, security, computer

4) The term computer _______________ is used frequently, but the _______________ of a computer is vulnerable to few risks unless the computer is connected to other computers on a network. As the use of computer networks, especially the Internet, has become pervasive, the concept of _______________ security has expanded to denote issues pertaining to the networked use of computers and their resources.

content, security, computer

5) In an email message, an intruder can easily fabricate where the email came from. But this information forging - called spoofing by intruders and _______________ professionals - is not limited to just email. In fact, the basic unit of _______________ transferred on the _______________- called a packet - can also be easily forged or spoofed.

internet, information, security

6) Think of your ________________ as you would your house, your apartment, or your condo. What do you know about how that living space works, what do you routinely do to keep it ________________, and what have you installed to improve its ________________?

security, computer, secure

7) The ________________ of computer security are based on ________________. As ________________ is not necessarily the primary goal of most computer applications, designing a program with security in mind often imposes restrictions on that program's behavior.

technologies, security, logic

8) While confidentiality, integrity, and authenticity are the most important concerns of a computer ________________ manager, privacy is perhaps the most important aspect of computer security for everyday Internet users. Although users may feel that they have nothing to hide when they are registering with an Internet site or service, ________________ on the Internet is about protecting one's personal information, even if the information does not seem sensitive. Because of the ease with which information in electronic format can be shared among companies, and because small pieces of related information from different sources can be easily linked together to form a composite of, for example, a person's information seeking habits, it is now very important that individuals are able to maintain ________________ over what information is collected about them, how it is used, who may use it, and what purpose it is used for.

control, privacy, security

9) ________________ security policies underpin the ________________ and well being of information resources. They are the foundation, the bottom line, of information security within an ________________.

organization, security, information

10) There are various ______________ and techniques used to design security systems. However there are few, if any, effective strategies to enhance ______________ after design. One technique enforces the principle of least privilege to great extent, where an entity has only the privileges that are needed for its function. That way even if an attacker gains access to one part of the ______________, fine-grained security ensures that it is just as difficult for them to access the rest.

strategies, security, system

13

English for Presentations

13 English for Presentations

13.1 프레젠테이션에서 자주 사용되는 영어 표현

인사법

발표를 시작할 때의 감사의 표현으로 시작하게 된다.

I am glad to ~

I am pleased to ~

I am happy to ~

서론의 전개

감사의 표현과 간단한 자기 소개를 한 후에 발표하고자 하는 주제에 대한 목적을 설명한다.

The purpose of my presentation is ~

I'd like to give a talk about ~

이 때 서론은 되도록 간략하게 준비한다. 전체 발표의 10~20% 정도로 준비하는 것이 바람직하다.

본문의 시작

본문은 이야기하려고 하는 요지를 처음으로 들어내는 부분이다. 따라서, 다음과 같은 구문으로 시작하는 경우가 많다.

To begin with ~

To start with ~

I'd like to begin by ~ing

본문에서 논평이나 비평으로 의견을 제시할 때는 'comment'라는 명사를 사용하는 것이 좋으며 가벼운 감상이나 의견을 제시할 때는 'remark'라는 명사를 사용하는 것이 좋다.

Let me make some comments on ~

I'd like to make one remark about ~
주장을 하나 하나 펴 나갈 때에는 다음과 같은 표현을 사용한다.
그 첫 번째로 First, 또는 First of all,
두 번째라면, Second,
마지막 주장이라면, Finally, 라고 표현한다.

발표 자료를 사용할 때의 표현

발표 자료를 슬라이드라고 부르는데, 발표를 할 때 슬라이드를 봐 달라는 영어 표현을 종종 들어 볼 수 있다. 표현은 다음과 같다.
Let's look at the next slide.

그래픽 개체 표현

또한 발표 자료는 시각화 된 객체의 표현이 많다. 특히 그래프 사용이 많아 그래프 설명에 대한 표현도 알아 보자.
증가하는 표현을 나타내는 단어는 다음과 같다.
increase, rise, go up, ascend, jump, grow

감소를 표현하는 단어는 다음과 같다.
decrease, fall, go down, descend, drop

변화를 나타낼 때는 vary, change, alter 등으로 표현한다.
더 세분화 해서 remain stable, stabilized, level out, fluctuate, move up and down, reach a peak, converged, remain constant, touch the bottom 등으로도 표현된다.

이와 함께 사용되는 부사로는 slightly, steadily, substantially, considerably, sharply, rapidly, dramatically, marginally, fractionally, somewhat 등으로 이런 부사를 사용함으로써 발표 자료의 내용을 더 구체화 시킬 수 있다.

경향 표현

경향을 표현하는 경우도 다양하다.
tend to ~

have a tendency to ~

원인과 결과 표현

원인으로부터 결과를 나타낼 때는 (원인) result in (결과), lead to, bring about, cause 등으로 표현하고, 결과에서 원인을 표현할 때는 (결과) results from (원인), be caused by, be due to, be produced by, be the effect of 등으로 표현한다.

구성의 표현

어떤 사물의 구성 요소를 표현하고자 할 때는 다음과 같은 표현을 주로 사용한다.

consist of ~

be made up of ~

be composed of ~

비교의 표현

자료의 동등함이나 유사함을 나타내고자 할 때는 다음과 같은 표현을 사용한다.

equal

be equal to ~

be identical to ~

be the same as ~

be similar to ~

be like ~

반면에 상이함을 나타낼 때는 다음과 같은 표현을 사용한다.

be different from

be distinguished from

contrast with

결론

마지막으로 결론으로 발표를 마무리하고자 할 때의 표현을 알아보자.

so, thus, hence, therefore, as a result, as a consequence, accordingly 등으로 표현한다.

그리고, 반드시 맺음말을 해서 발표한 내용을 다시 한번 정리하는 것이 좋다.

I'd like to summarize my talk ~

I'd like to make some conclusions.
Finally, I can conclude that ~
This wraps up my presentation.
This is all I have to say today.
In conclusion ~

질의 응답

발표를 갑작스럽게 종료하는 것은 무례한 행동이다. 따라서, 반드시 질의 응답 시간을 갖도록 하자. 이때 사용하는 표현은 다음과 같다.

May I ask you a question?
I want to ask you some questions.
There is something I want to ask.
I have a question about ~
I am wondering ~
I am interested in ~
I'd like to hear ~
I am curious that ~
I am wondering whether ~

응답 표현으로는 다음과 같다.

sure, surely I can, certainly 등으로 시작해 보자.
I'd like to answer that question.
My answer to the question is that ~
I will respond to that question briefly.

13.2 수식과 기호의 영어 표현

소수 표현

1 one tenth (point one)
01 one hundredth (point zero one)
001 one thousandth (point zero zero one, point two zeros one, point two Os one)
8.261 eight point two six one

0.743 zero point seven four three, zero decimal seven four three

분수 표현

2/3 two over three, two-thirds, two divided by three

가감승제 표현

2 + 3 = 5

Two plus three equals five, Two and three is five, Two added to three makes five

3 −2 = 1

Three minus two equals one, Two from three leaves one, Two from three is one

A * B

to multiply A by B, the product of A and B

6 / 2 = 3

Six divided by two equals three

좌표에 관한 표현

좌표 a coordinate

x-axis, y-axis

원점 the origin

1차 방정식 a first degree equation, a linear equation

상수 constant term

경계가 있는 bounded

the upper bound, the lower bound

극한치 the limit value

무한대 infinity

이외에도 여러 가지의 수의 관한 표현으로 실수, 유리수, 무리수, 정수, 홀수, 짝수를 표현하는 방법과 벡터나 스칼라를 표현하는 것도 공학도들에게는 필수이다.

a real number, a rational number, an irrational number (complex number), an integer, a fraction, an odd number, an even number

the component of a vector x, the cell in the i row, the j column

50 가지 파워포인트 표현에 대한 팁

01. Good afternoon. I'm happy to introduce our new line of products today.
안녕하십니까. 오늘 우리의 신제품을 소개하게 되어 기쁩니다.

02. My name is John Walters, and I'm in charge of product design.
제 이름은 John Walters이며, 제품디자인을 담당하고 있습니다.

03. Let me explain the four musts of choosing a good overseas distributor.
좋은 해외 공급자 선택 시 필수 사항을 네 가지 말씀 드리겠습니다.

04. I've divided my talk into three parts: First, the overview of the project, then the financial aspects, and finally our projections.
말씀드릴 내용은 크게 3가지입니다. 첫째 프로젝트 개요, 둘째 재정적인 측면, 마지막으로 우리의 전망입니다.

05. This presentation will take about 20 minutes.
이 프레젠테이션은 20분 정도 걸릴 것입니다.

06. Please interrupt me at any time if you have any questions.
도중에 질문이 있으면 언제라도 말씀해주십시오.

07. I'm going to start with a brief overview of our core business.
우리의 핵심사업을 간단히 개관하는 것으로 시작하겠습니다.

08. Now that we've covered the advantages, ...
지금까지 이점들을 짚어보았고...

09. Let's move on to the next part.
다음 장으로 넘어가겠습니다.

10. My final topic is on increasing output.
마지막으로 드릴 말씀은 생산량 증대에 관한 것입니다.

11. I'll briefly summarize the main points of the takeover offer.
그 인수제안의 요즘들을 간략히 정리해 보겠습니다.

12. So that concludes the financial aspects of the project.
이것으로 프로젝트의 재정적인 측면을 마치겠습니다.

13. Before I wrap up my talk, are there any questions about our payment structure?
이야기를 마무리하기 전에, 우리의 지불 구조에 대해 질문 있으십니까?

14. In closing, I would like to propose the following.
끝으로 다음과 같은 제안을 드리고 싶습니다.

15. According to the latest reports, we hold 25% of the auto market.
최신보고서에 따르면, 우리는 자동차 시장의 25%를 점유하고 있습니다.

16. A good example of that is Japan's real estate bubble bursting.
일본의 부동산 거품 붕괴가 그 좋은 예입니다.

17. Experts estimate that crude oil prices could reach $50 a barrel.
전문가들은 원유가격이 배럴당 50달러에 이를 것으로 예측하고 있습니다.

18. If you compare their services and ours, you can see why we're better.
그들의 서비스와 우리의 서비스를 비교해 보면, 왜 우리가 더 나은지 알 수 있을 겁니다.

19. Let me illustrate this point with some hard facts and figures.
이 점에 대해 몇 가지 실제사실과 수치로 설명 해 보겠습니다.

20. Please take a look at the pie chart on the screen.
스크린에 있는 파이차트를 봐주십시오.

21. As you can see from this graph, our profit's been increasing steadily.
이 그래프에서 보듯, 우리의 수익은 꾸준히 증가하고 있습니다.

22. The bar graph on the right shows our export revenues.
오른쪽 막대 그래프는 우리의 수출 총수입을 보여주고 있습니다.

23. The vertical line represents the market share, and the horizontal line represents sales volume.
세로선은 시장 점유율을 나타내고, 가로선은 판매량을 나타냅니다.

24. The next step will be finding appropriate construction sites.
다음 단계는 적합한 건축 부지를 찾는 것입니다.

25. By the way, you may have seen articles about this project.
그런데, 아마 이 프로젝트에 대한 기사를 보신 적이 있을 겁니다.

26. This survey was based on interviews with 500 executives.
이 조사는 경영간부 5백 명의 인터뷰를 바탕으로 이루어졌습니다.

27. If you look at this chart, you'll see our target markets.
이 차트를 보면 우리의 주력시장을 알 수 있습니다.

28. On the other hand, their beers are much more bitter than ours.
반면, 그들의 맥주는 우리보다 맛이 훨씬 더 씁니다.

29. Depending on how the customers react, there are a few steps we can take.
고객들의 반응이 어떤가에 따라, 우리가 택할 수 있는 단계들이 몇 가지 있습니다.

30. There are three things to consider in choosing the site; first, transportation, second, labor costs, and third, local and state taxes.
장소를 고를 때 고려해야 할 점이 세 가지 있습니다. 첫째 교통, 둘째 인건비, 셋째 지방세와 주세입니다.

31. Contrary to popular belief, we expect significant growth in our industry.
통념과는 반대로, 우리는 우리 업계에서의 눈부신 성장을 기대하고 있습니다.

32. There are many differences between their product and ours.
그들의 제품과 우리의 제품에는 많은 차이점이 있습니다.

33. That means we can expect our market share to improve.
다시 말해, 우리의 시장점유율 상승을 기대할 수 있다는 뜻입니다.

34. We should pay close attention to interest movements.
우린 이자의 변동에 세심한 주의를 기울여야 합니다.

35. As a result, we were able to grow 25% annually.
결과적으로 우린 매년 25% 성장 할 수 있었습니다.

36. We expect demand to increase 50% over the next few years.
우리는 수요가 몇 년 안에 50% 증가할 것으로 기대합니다.

37. We're forecasting a significant decrease in our exports to South America.
남미로의 수출은 크게 감소할 것으로 전망됩니다.

38. Our agricultural products account for 25% of our sales.
우리의 농산품은 판매의 25%를 차지합니다.

39. Incredibly, our profits grew two-fold annually for the past ten years.
놀랍게도 우리의 이익은 지난 10년간 매년 두 배 증가 했습니다.

40. Consumer spending fell drastically last year.
소비지출이 지난해에는 현저히 줄었습니다.

41. We are a leading biochemical products manufacturer in Asia.
우리는 아시아에서 손꼽히는 생화학제품 제조업체입니다.

42. Korea Tires was established in 1995.
Korea Tires는 1995년에 설립되었습니다.

43. We specialize in educating and training English teachers.
우리는 영어교사 교육과 양성을 전문으로 합니다.

44. We have been in this industry for more than 10 years.
우리는 이 산업에 10년 넘게 종사해왔습니다.

45. K Corp. is committed to providing the highest quality high-end products for women.
K 법인은 최상급 품질의 최고급 여성용 제품을 제공하는 데 전념하고 있습니다.

46. The Korea Group consists of seven autonomous companies.
Korea 그룹은 7개의 독립계열사들로 이루어져 있습니다.

47. Our revenue reached $575 million in 2015-a 45% increase over the past year.
우리의 순이익은 2015년에 5억 7천 5백만 달러에 도달하여, 전년도 대비 45% 증가했습니다.

48. We expect our revenue to grow by 27% annually.
우리의 총수입은 매년 27%씩 신장될 것으로 기대됩니다.

49. Korea Group's goal is to become the leading tire manufacturer.
Korea 그룹의 목표는 정상급 타이어 제조업체가 되는 것입니다.

50. We are planning to expand into North American markets in the near future.
우리는 조만간 북미시장에 진출할 계획입니다.

References [참고문헌]

[1] 융합 그리고 넥스트, 손병희 저, 교우사, 2015
[2] Cambridge Advanced Learner's Dictionary, Cambridge University Press
[3] Collins COBUILD Intermediate Learner's Dictionary, Heinle & Heinle Publishers
[4] Longman Essential Activator, Longman National Corpus
[5] English Grammar in Use, Raymond Murphy, Cambridge University Press

13과 연습문제

01 **다음을 영작하여 표현해 보시오.**

- 오늘 우리회사의 신제품에 대해 소개하게 되어 기쁩니다.
- 이야기를 마무리하기 전에, 제 발표에 대해 질문 있으십니까?

1) 다음 그림에서 그래프의 변화를 표현해 보시오.

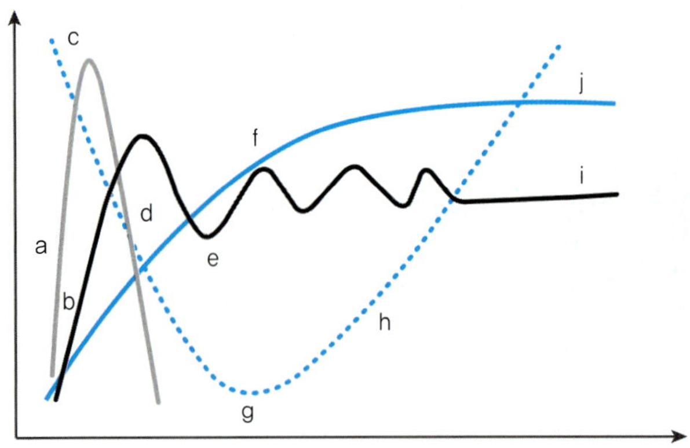

14

Preparing Presentations

14 Preparing Presentations

14.1. Notes to the Presenters

1. When presenting a paper it is essential to consider the type of audience you will be addressing. Who are they? Are they experts in your field, scientists with an interest in the topic or non-technical people who need to understand the general idea of what you are doing.
2. Having determined your audience the next step is to decide what you want to tell them. In planning your presentation you must first answer the question '*why do I want to talk to these people?*' Are you presenting scientific findings in order to persuade the audience to make a policy decision, to review your research grant or simply to explain your research to colleagues?
3. The purpose of a presentation is to make the audience want to understand more about your subject, and in turn want to discover more afterwards. For example, at a conference, you should assume that people have not read your paper (which is normally the case), so *you should try to make them want to read it.* Remember that at many conferences, there are many people prepared to provide support for your research; show them that you can deliver the products. Remember always that *the audience will not know the subject as well as you.*
4. Under no circumstances should you read your paper. Even if English is not your first language it is always a mistake to read.
5. Remember always that people have paid to attend the conference and that they will complain to the organizers if they do not find the event worthwhile. This will result in less support for the conference and less opportunities for you to present your work in the future. *The success of the meeting depends on you.*
6. You may wish to improve your presentation by encouraging the audience to ask questions during your presentation. If you decide to do so, please ask permission from the Session Chairmen as sometimes this will not be possible as it may disrupt the session schedule. You should be careful in these cases

not to be sidetracked or overrun the time allocated for your paper.

When Presenting

7. Structure your presentation in a similar way to your written paper. First introduce yourself and the presentation, then move to the main body of the paper. This should start by describing the problem to be addressed, then explain the technique used and follow with the results of your study. Having done that, draw your conclusions and describe future work if relevant. *Stress to your audience the significance of your work*, the new developments that you have discovered and the main aspects of your work.
8. Practice your presentation in front of people who do not understand your work.
9. *Be sure that you keep to the schedule* and always allow 3 – 5 minutes for questions and discussions. The timing is important as you may be cut off by a forceful Chairman before you have presented your results and conclusions.

Visual Aids

10. *Your presentation will be greatly enhanced with the use of good visual aids.* Your aim should be to make your presentation as easy to follow as possible.
    - Try to use landscape format where possible which is similar to the human vision. Avoid vertical slides and transparencies.
    - Use colour wherever possible.
    - Put your organization's name at the bottom corners of each slide and maybe your own name.
    - The first slide should have the name of your presentation, your name and possibly the conference name and date.
    - Do not use too many equations. The audience can always read them in the paper if necessary.
    - Have between 3 – 5 points per slide. More than 5 is too many and you should use a second slide.
    - Allow 2 – 3 minutes per slide and remember to pause so the audience can read the whole slide. Do not block part of the slide or transparency (i.e. by standing in front of the projector or by design (if you wish to present parts of it later on).

- Remember that a picture or graph is very informative. Check beforehand that your visual aids are of good quality and that they can be read from a distance.
- Turn the slide and overhead projectors off if they are not being used for your presentation.
- Use the pointer as much as possible and always try to face the audience as much as possible.

11. Let the organizers know well in advance if you are planning to use other visual aids such as video tapes or computer display terminals. In some cases if they are told in advance they may be able to provide better facilities.
12. Also let the organizers know well in advance if you will require to screen simultaneously or to use special projection facilities. Do not assume that they will be available as a matter of course or that they can be found at short notice. Make sure your visual aids are coordinated with the conference visual aids well before the session. Meet your session Chairman at least 10 minutes before the session starts.

General Teps for Making Your Owm Presentation

- 프레젠테이션은 일정한 형태의 제안이나 아이디어를 발표하는 것을 의미한다.
- 프레젠테이션은 시간과 장소의 제약을 받는 데다 1회로 끝나는 경우가 대부분이므로 글쓰기의 - 기술과 말하기의 기술의 결합으로 아주 어려운 작업이라 할 수 있다.
- 글쓰기의 기술이라면 발표 자료를 만드는 것을 의미하고, 말하기 기술이라면 발표할 때의 발표자의 스킬 이라고 말할 수 있다.

자료의 작성 순서는 다음과 같다.

1. 발표 주제 설정
2. 자료 수집과 검토
3. 전체 전략 수립
4. 세부 내용 작성
5. 디자인 및 효과 점검

글 쓰는 능력을 향상시키는 방법으로 다음과 같은 제안을 제시해 본다. 좋다고 생각되는 글을 직접 적어보고 그 글에 대한 자신의 생각이나 느낌을 직접 적어 보자.

14.2. Instructions for PowerPoint

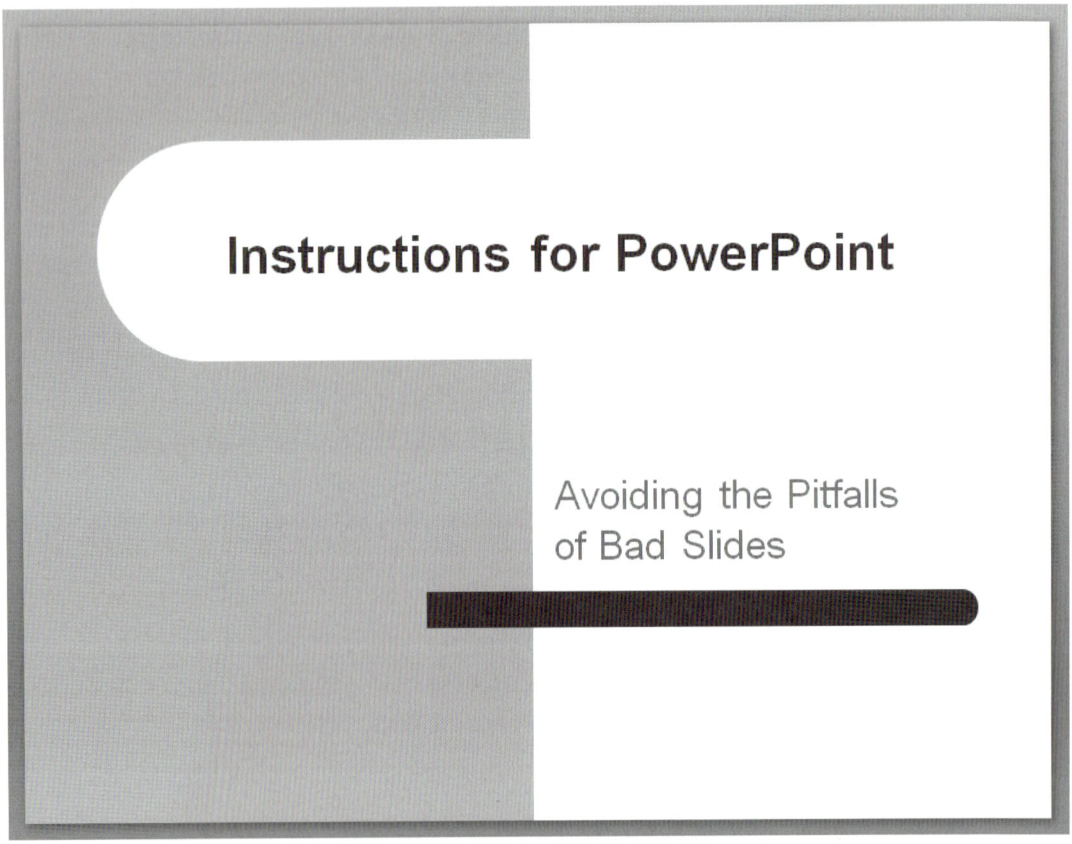

Comments

다음은 발표자료를 만드는 좋은 팁을 소개하고자 한다.

- 발표 자료는 파워포인트라는 소프트웨어를 이용해서 주로 만들게 되는데, 파워포인트에서는 각 페이지를 슬라이드라고 부른다.
- 먼저 발표 자료에서 가장 첫 번째로 오는 슬라이드는 제목 슬라이드로 이 슬라이드에는 발표 자료의 제목을 적게 된다.
- 제목을 적은 뒤엔 부제목으로 발표자의 소속과 발표자의 이름, 그리고, 발표자의 이메일 주소를 적어 주는 것이 형식을 갖춘 제목 슬라이드이다.

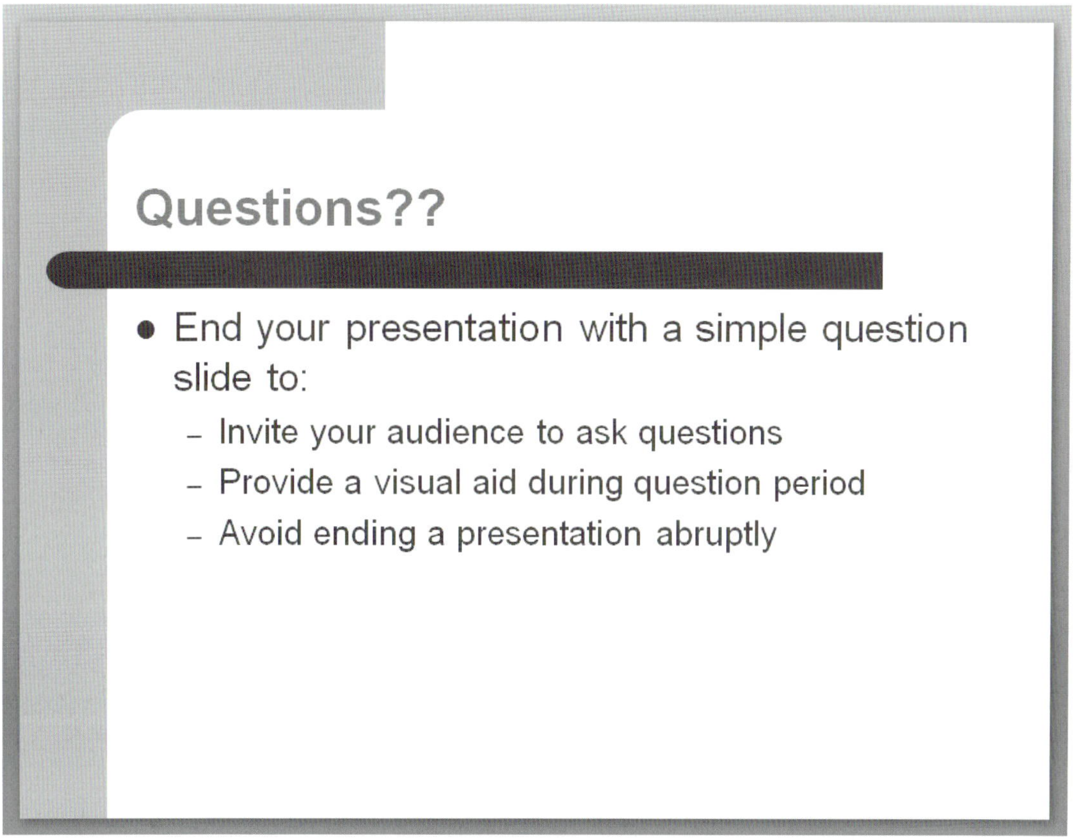

Comments

- 발표 자료의 가장 마지막 슬라이드로는 발표를 마치고 질의 응답하는 시간에 띄어 놓을 수 있는 간단한 질의 응답 슬라이드이다.
- 발표를 할 때 갑작스럽게 슬라이드를 종료하는 것은 피하기 위해서 이다.

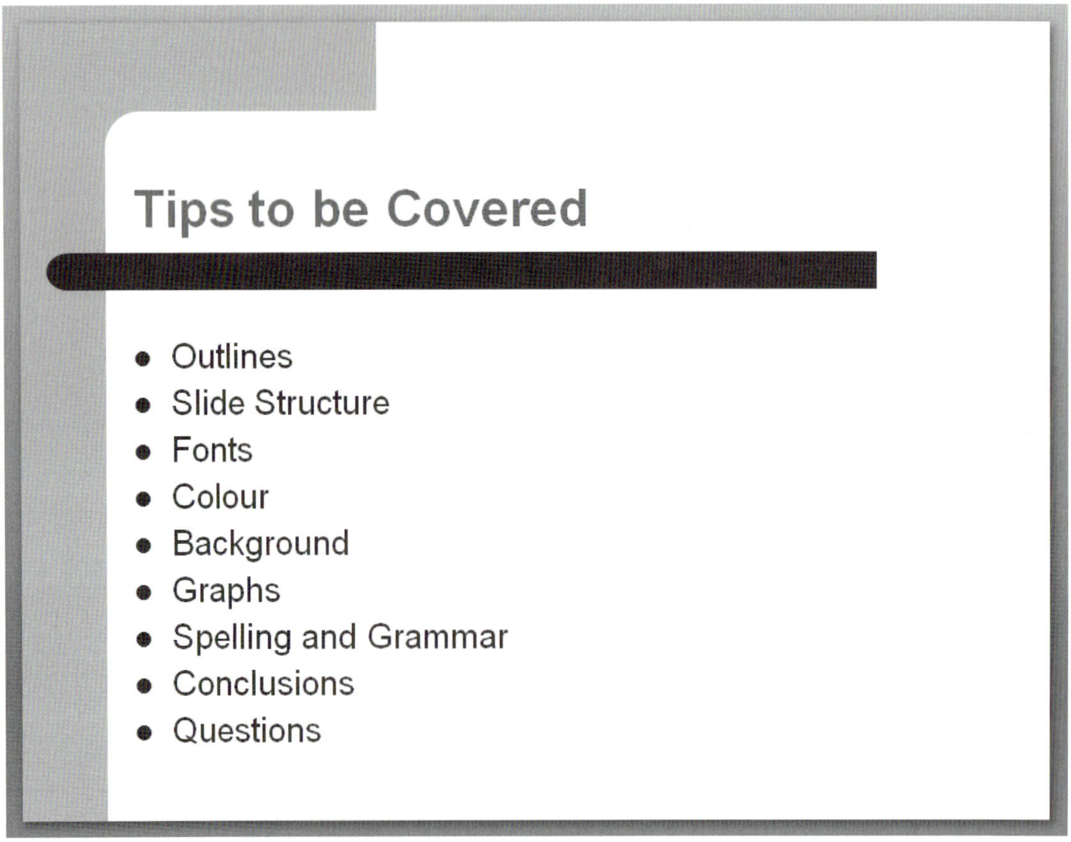

Comments

- 자료 작성할 때의 방법을 정리해 보자.
- 나열식의 설명보다 요점을 정리한다.
- 요점을 정리할 때 그래프나 도형 개체 등을 활용하여 시각화 시키는 것이 가장 좋다. '~하다' 등의 서술형보다는 명사형으로 본문을 만든다.
- 이/가, 을/를 등의 조사는 꼭 필요한 경우에만 사용한다.
- 마침표나 느낌표 등 문장 부호는 가능한 한 사용하지 않는 것이 좋다.
- 누구나 알기 쉽게 평이한 어휘를 구사하는 것이 좋다.
- 설명이 필요한 용어는 말 풍선 효과를 사용한다.
- 어휘 중복은 피하되 강조해야 할 단어는 여러 번 반복해서 사용되는 것은 허용한다.

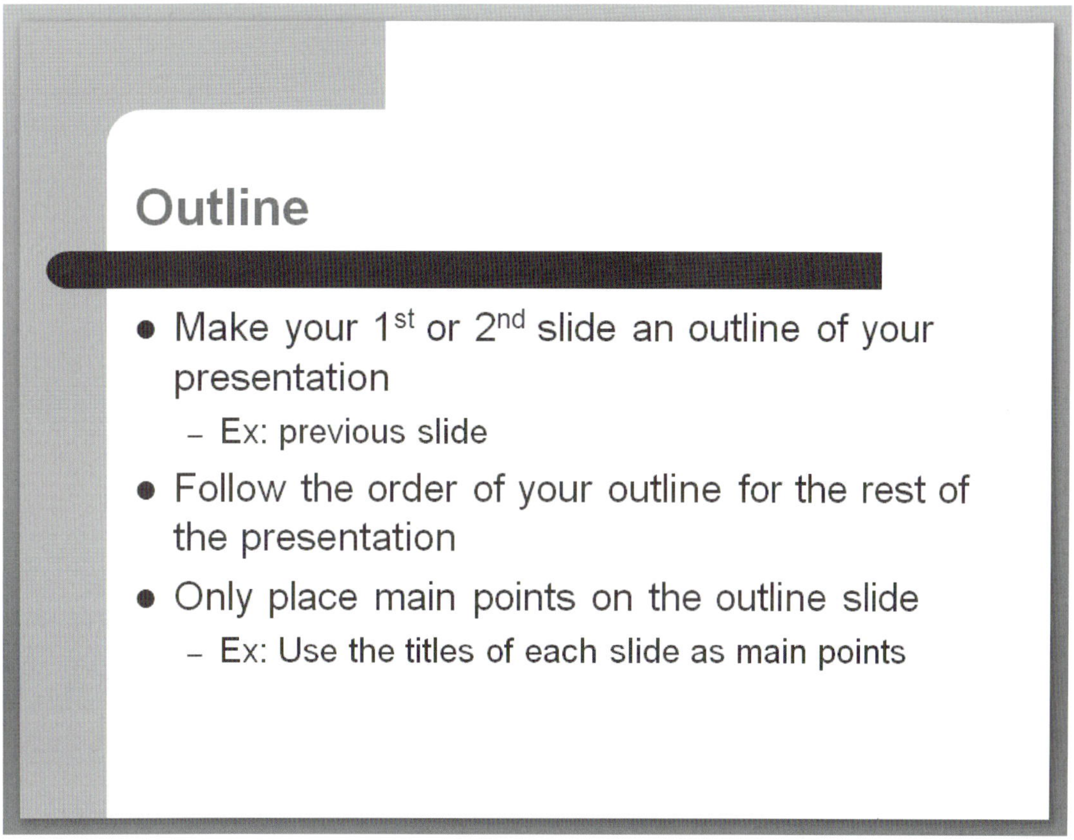

Comments

- 제목 슬라이드 다음에는 목차를 만든다. 목차는 각 슬라이드의 제목으로 한다. 따라서, 발표자료를 만드는 순서를 정리해 보면 다음과 같다.
- 제목 슬라이드 작성
- 본문 슬라이드 작성
- 목차 작성
- 질의 응답 슬라이드 작성

위의 순서로 만들되 목차는 제목 슬라이드 다음에 위치 시킨다.

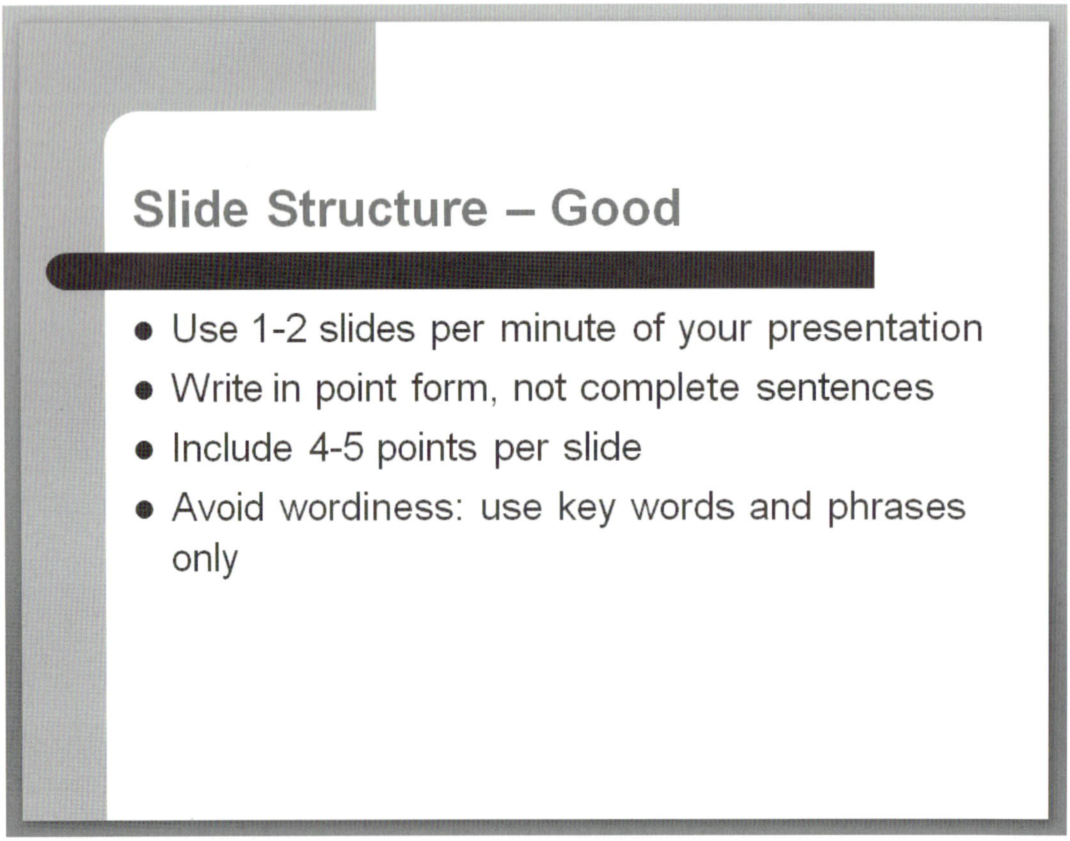

Comments

- 형식적 측면에서의 발표 자료 작성 방법에 대해 살펴 보자.
- 먼저, 화면 구성을 할 때는 기본적으로 가로형을 선택한다.

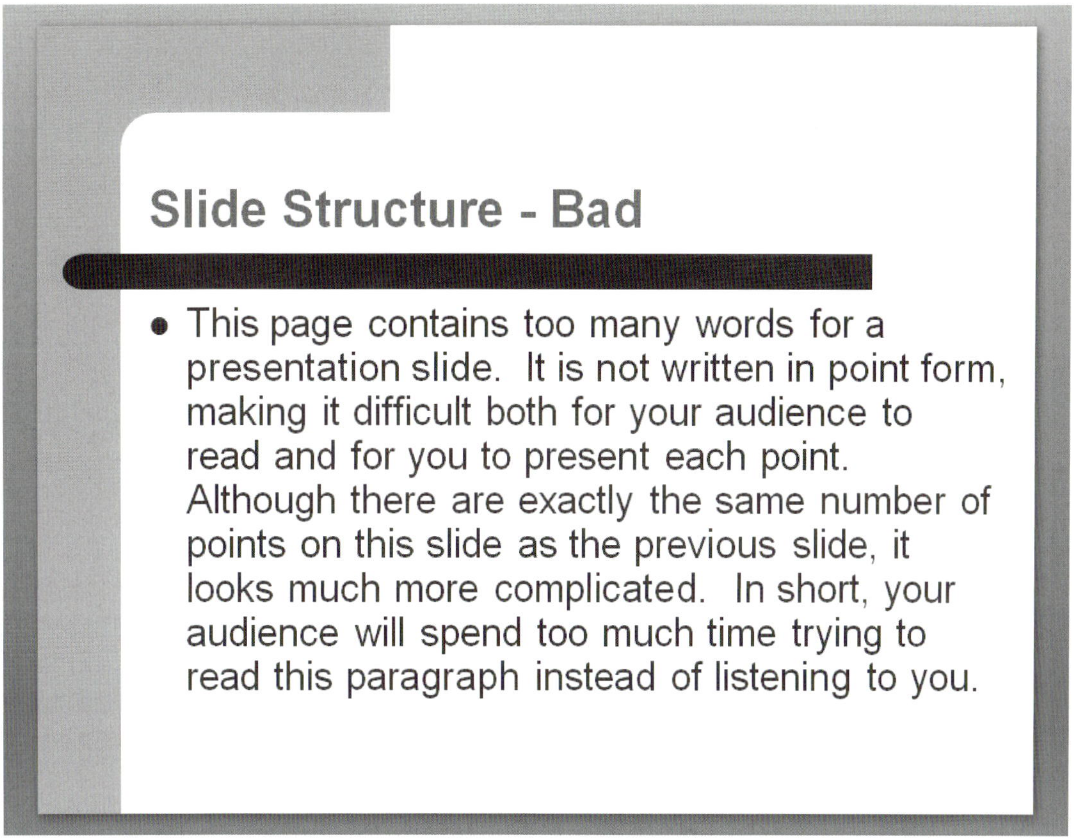

Comments

- 한 줄에는 하나의 정보만 표시한다. 단, 한 줄에 7단어를 넘지 않는다.
- 한 슬라이드에는 제목을 포함하여 6줄을 넘지 않는 것이 좋다.

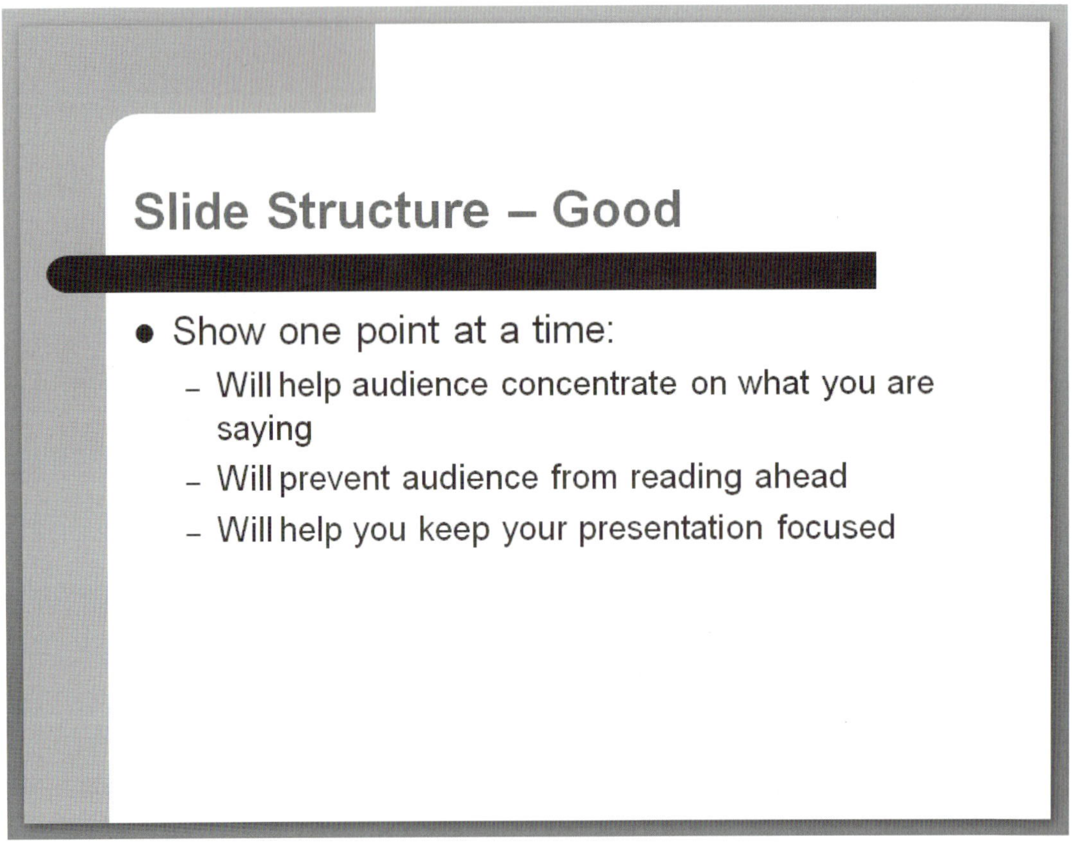

Comments

- 한 페이지에는 하나의 개념이나 핵심 정보만을 담는다.

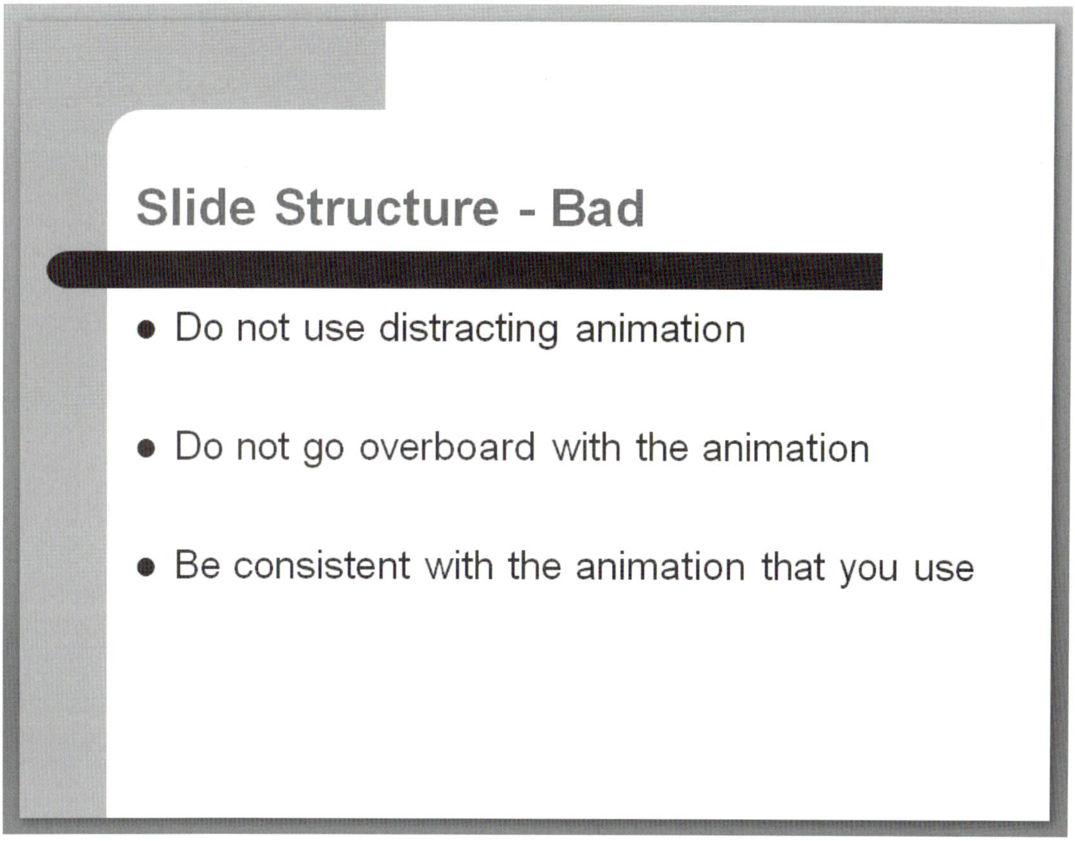

Comments

- 여백이 너무 많아도 좋지 않다.

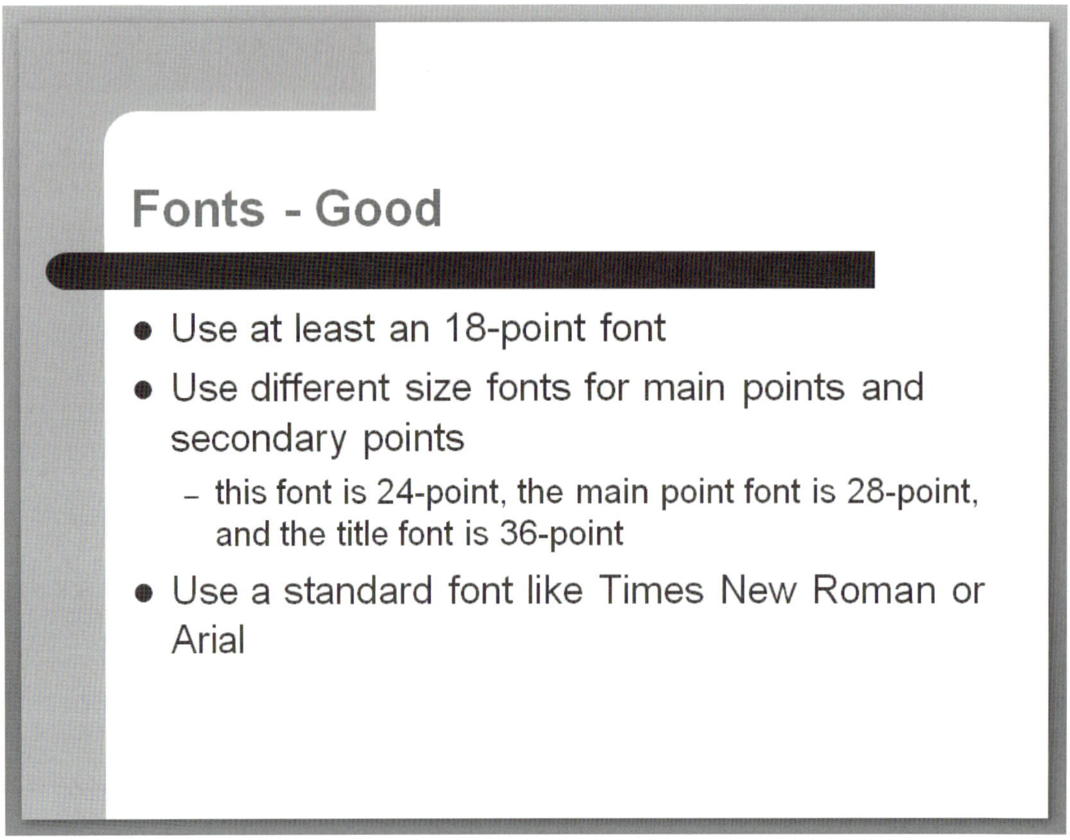

Comments

- 본문 텍스트 크기는 한글의 경우 24포인트가 좋으며 34포인트를 넘지 않는 것이 좋다. 제목 텍스트 크기는 36포인트에서 44 포인트가 적당하다.
- 본문 텍스트건 제목 텍스트건 너무 크거나 작으며 청중들의 시선을 집중시킬 수 없음을 꼭 기억하자.
- 제목 텍스트 글꼴로 가독성이 입증된 글꼴이 있다.
- 한글 글꼴로는 'HY헤드라인M'체로 굵게 표현되는 글꼴이 좋으며, 영문 글꼴로는 'Arial Black'과 같은 글꼴이 좋다. 본문 텍스트로 사용될 한글 글꼴로는 '굴림체, 바탕체, 맑은 고딕체' 등이 좋으며, 영문 글꼴은 'Times New Roman' 글꼴이 무난하다고 한다. 이런 글꼴은 여러 가지 글꼴을 동시에 사용하는 것보다 한 두 개로 사용하는 것이 바람직하다.

Comments

- 영문 자료를 만들 때 주의사항이라면, 대문자는 꼭 필요할 때만 사용하자.
- 대문자로 표기된 내용은 읽기가 어려울뿐더러 올바른 형식이 아니다.
- 대문자는 고유명사나 강조해야 할 때 사용한다.
- 제목을 입력할 때 관사나 전치사를 제외한 단어의 첫 글자는 대문자로 표현한다.

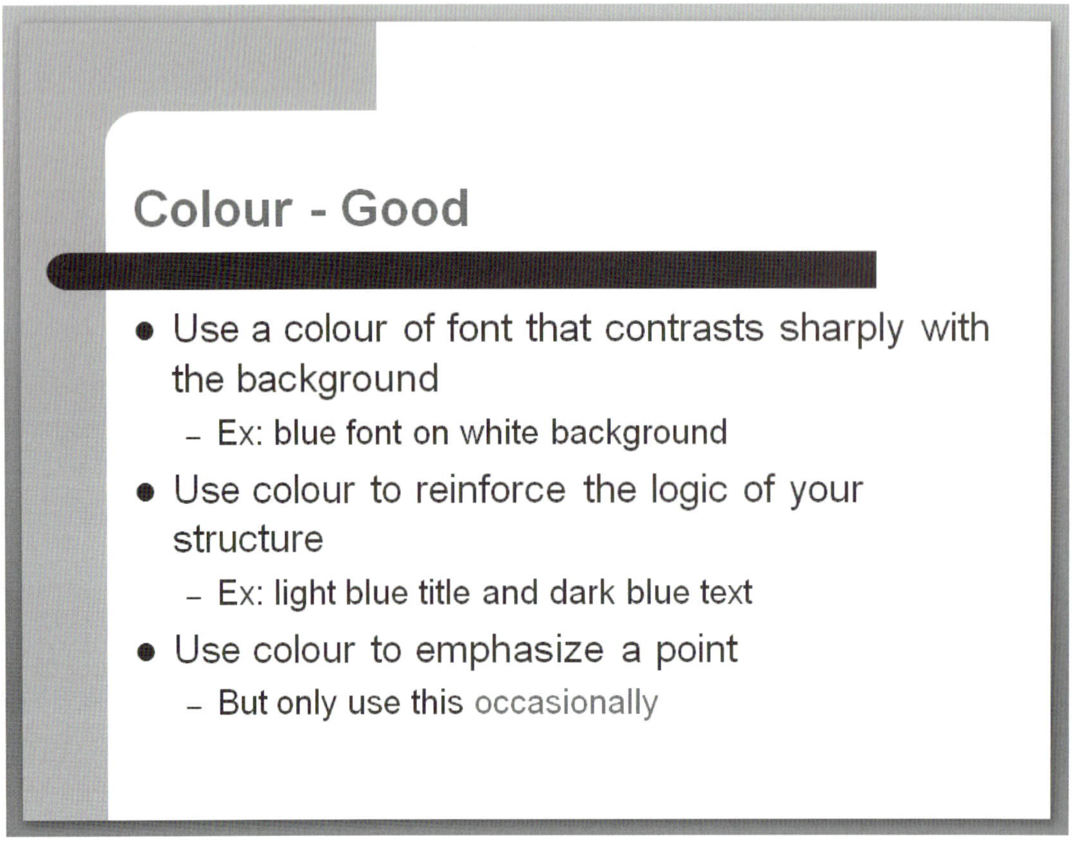

Comments

- 색깔은 검정을 포함하여 세 종류를 넘지 않는 것이 좋다.
- 색은 먼셀의 색 조합표를 참조해 보자.

  1943년 미국광학회 측색위원회(測色委員會)에서 수정하여 국제적으로 널리 사용되고 있는 것으로 색을 색상 (色相:H=hue)·명도 (明度:V=value)·채도 (彩度:C=chroma)의 삼속성 (三屬性)으로 나누어 HV/C라는 형식에 따라 번호로 표시하게 된다.

  먼저 색상환(色相環)을 10으로 나누고 필요에 따라 각각의 사이를 다시 반분한다. 즉 빨강(R)·노랑 (Y)·녹색 (G)·파랑 (B)·보라 (P)색을 기본 색으로 하고, 각각의 중간에 주황(YR)·연두 (GY)·청록 (BG)·남색 (PB)·자주 (RP)를 두어서 합계 10가지의 색상으로 나눈다. 그리고 각 색상 사이를 다시 또 10등분하여 번호를 붙인다. 이 분할에 따를 때 가장 빨강색다운 색상이 5R, 가장 녹색다운 색상은 5G가 된다.

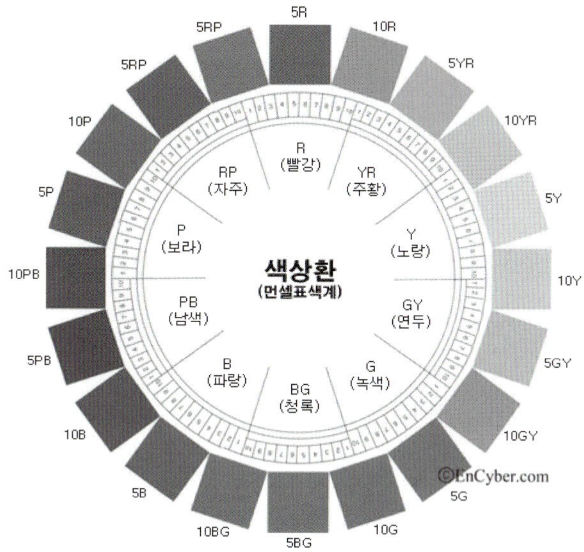

- 또 명도를 나타내는 척도로서는 흰색에서 흑색까지의 무채색(N)의 밝음을 등분하여 11단계로 하여 흰색을 10, 흑색을 0으로 하는데, 필요에 따라 이것을 다시 세분하여 소수점을 찍을 수도 있다. 채도의 척도로서는 무채색을 0으로 하고 그와 같은 감각 차에 준해서 순도가 높아짐에 따라 차례로 1, 2, 3,… 번호를 단다.

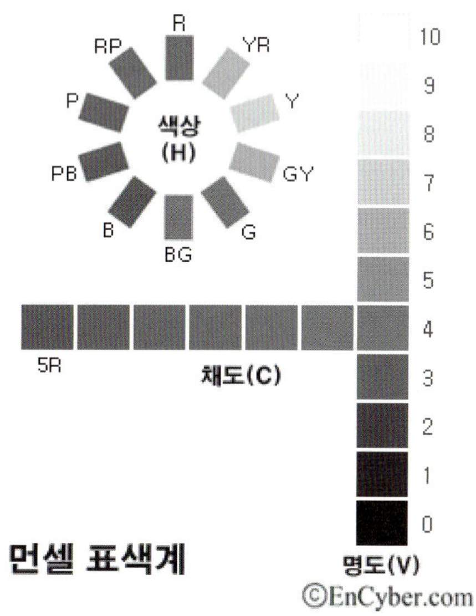

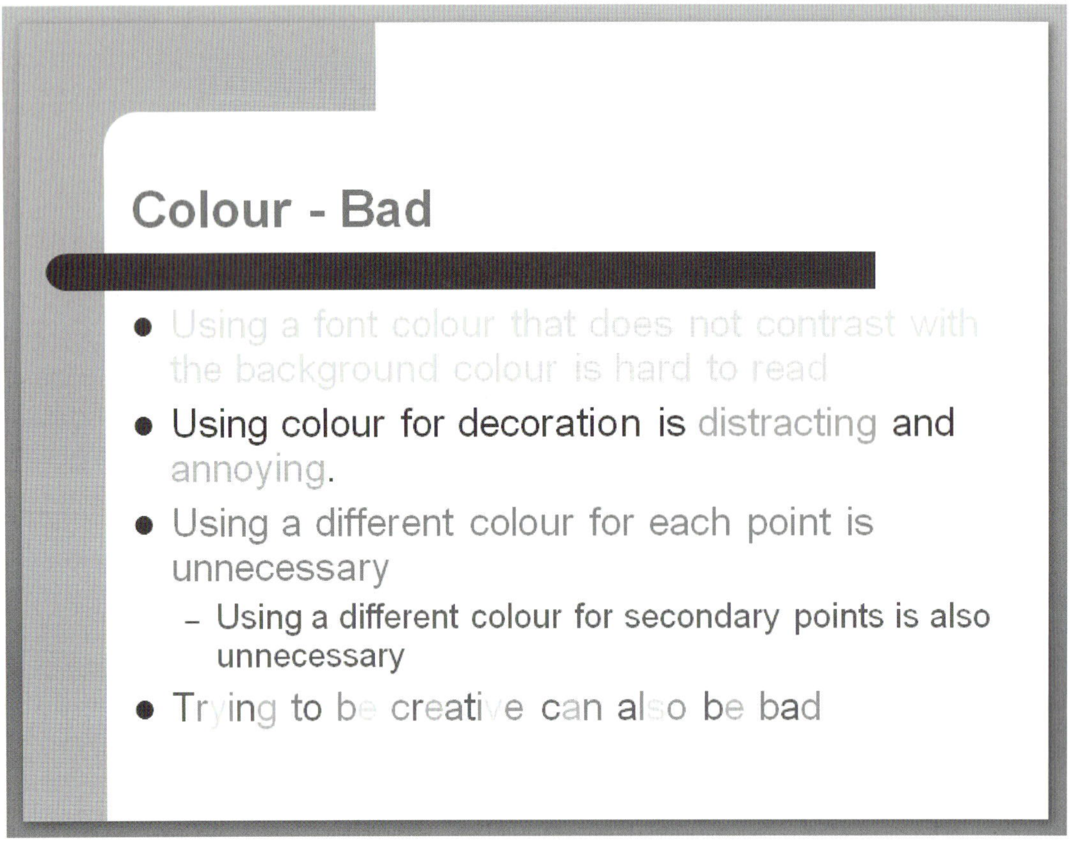

Comments

- 색은 유사 색으로 3가지 색을 이용한다면 무난한 느낌의 자료가 되고, 보색대비로 색을 이용한다면 화려함 느낌을 주는 자료로 만들 수 있으나 너무 많은 컬러를 사용하게 되면 오히려 복잡해 보이고 청중들에게 혼란만 줄 뿐이다.

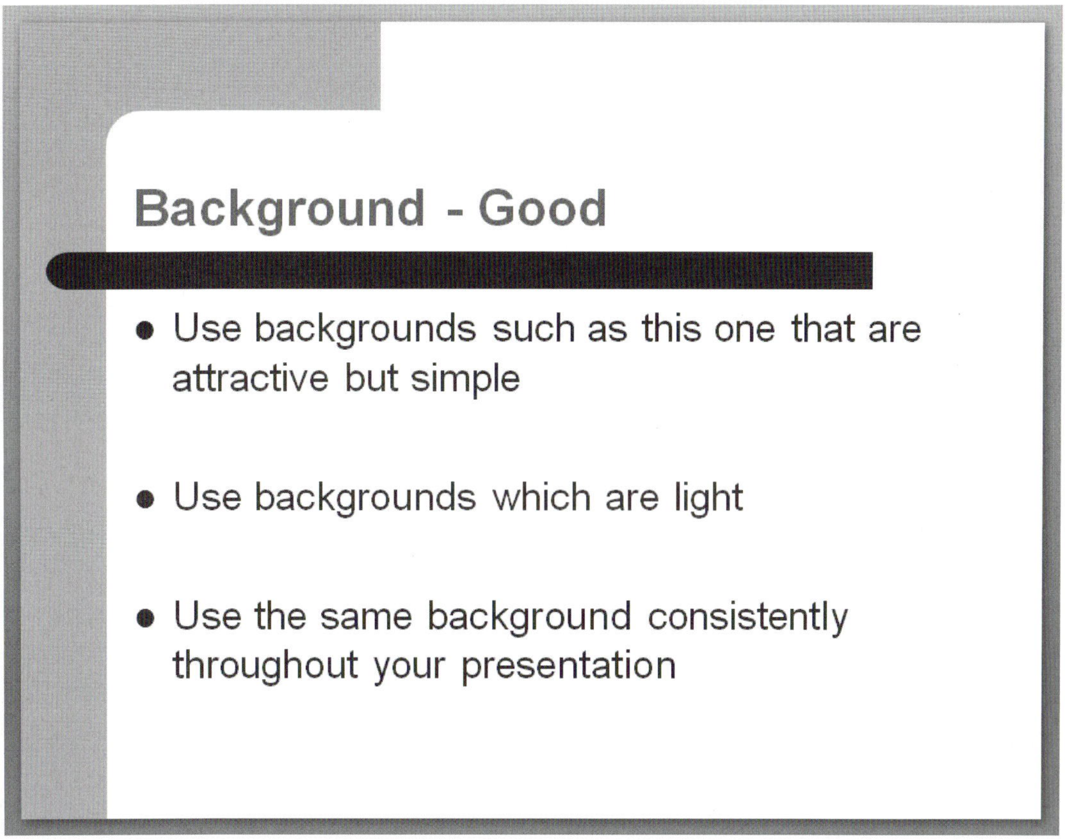

Comments

- 글꼴 색을 선택할 때 배경 색을 고려해서 선택해야 한다. 배경색이 어두운 계열이라면, 글꼴 색으로는 흰색이나 오렌지 계열의 색이 좋다. 반면 배경색이 흰색이라면, 글꼴 색으로는 검정색이나 어두운 청색계열의 색이 잘 어울릴 것이다.
- 슬라이드 장식은 전체적으로 통일감을 주고 일관성을 유지하며 단순한 것이 좋다.

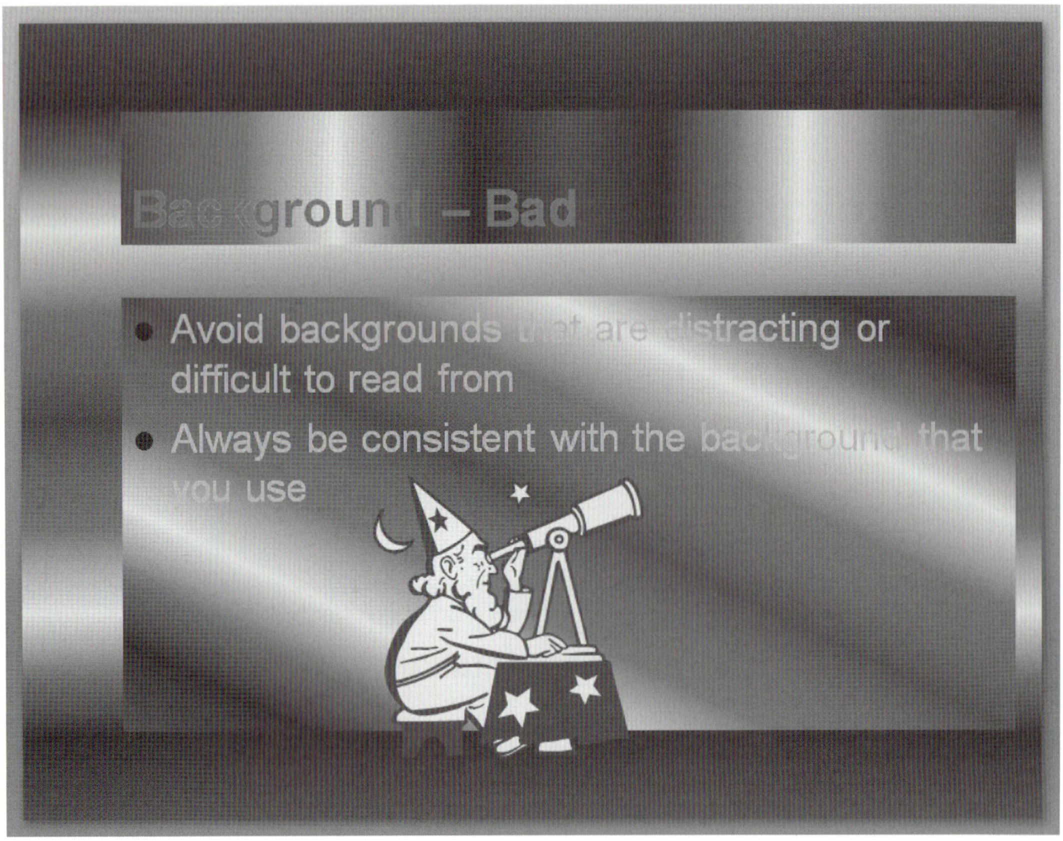

Comments

- 배경 색이 화려하면 오히려 역효과를 불러 올 수 있으니 각별히 주의하자.

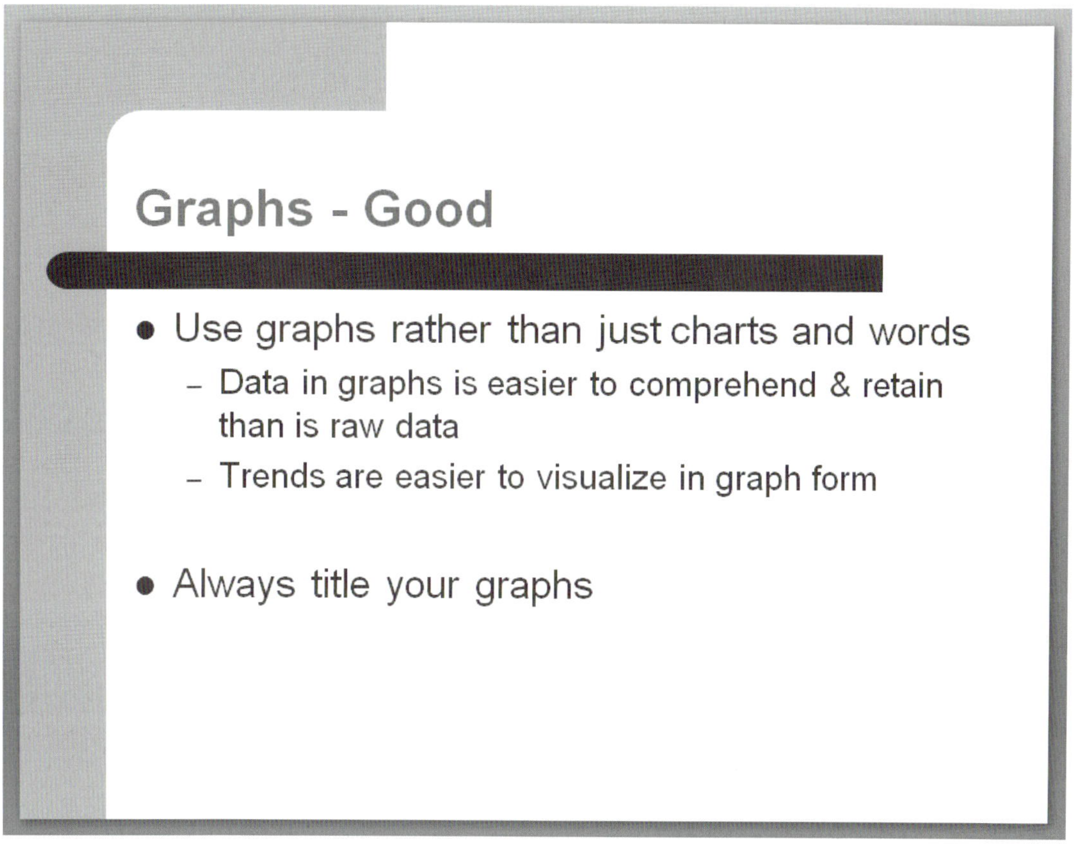

Comments

- 그래픽 개체를 사용하여 슬라이드의 내용을 시각화 시키는 것이 원칙이라고 생각하자.
- 그래픽 개체를 사용했다면, 그 출처나 단위, 그리고, 타이틀은 꼭 적어 주자.
- 이때 글자를 너무 작게 해서는 안 된다.

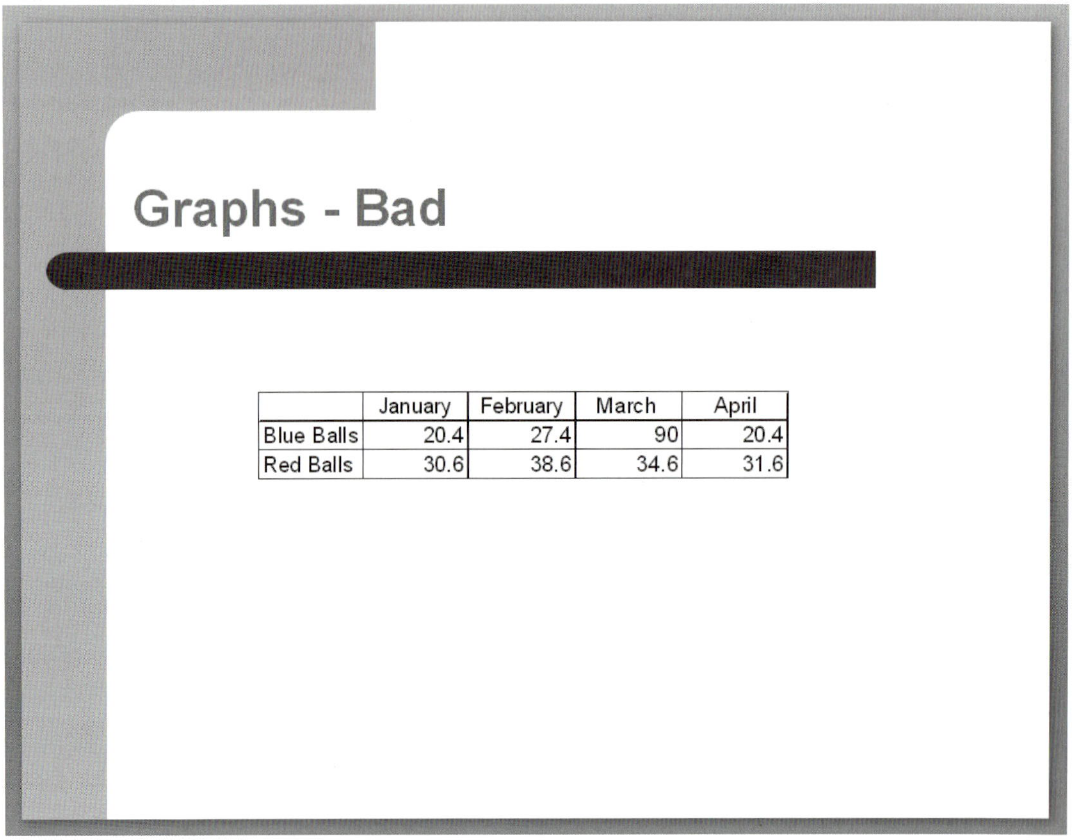

	January	February	March	April
Blue Balls	20.4	27.4	90	20.4
Red Balls	30.6	38.6	34.6	31.6

Comments

- 그래픽 개체를 사용할 때는 크기와 배치가 내용에 따라 리듬을 갖는지 꼭 살피자.
- 그래픽 개체로는 차트, 표, 도형, 조직도, 다이어그램, 그림, 사진 등으로 다양하다.
- 소리 개체나 동영상 개체도 활용할 수도 있다.

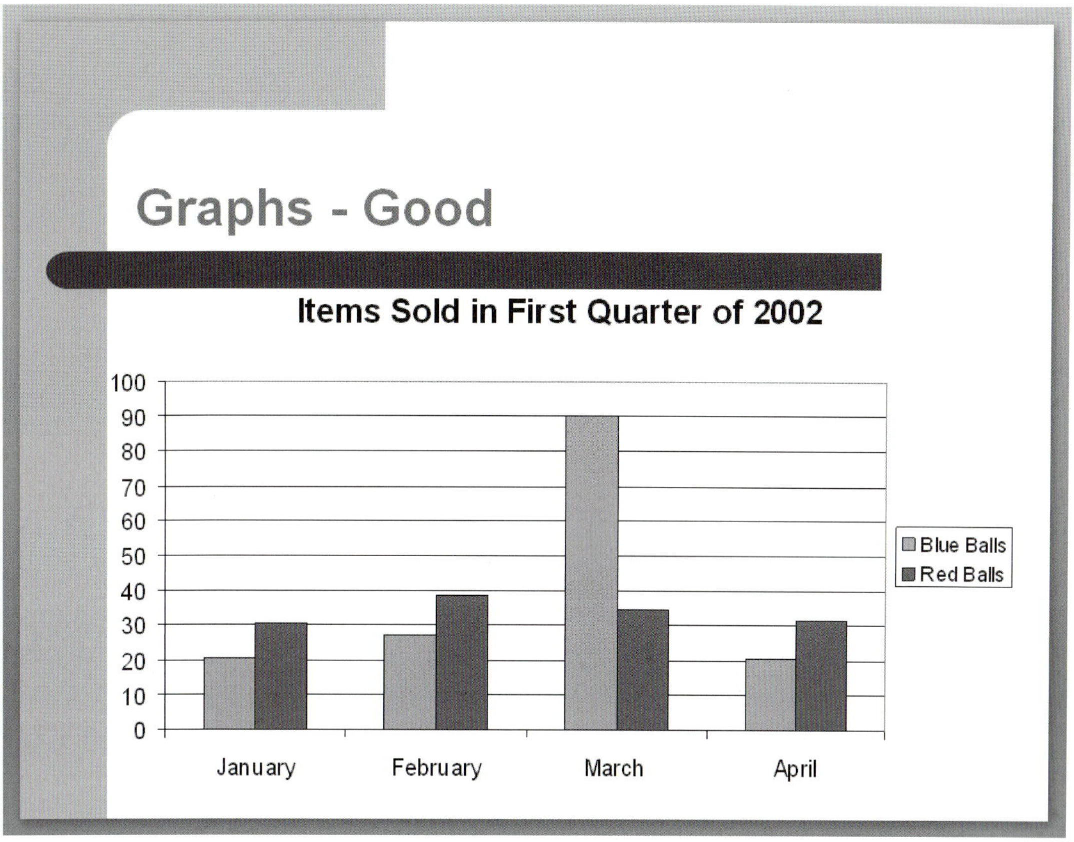

Comments

- 다양한 개체를 활용할 때도 전체 발표자료의 서식이나 리듬을 고려해야 하며 그래픽 개체만 계속 나온다든가, 글만 계속 나오는 것은 좋지 않다.

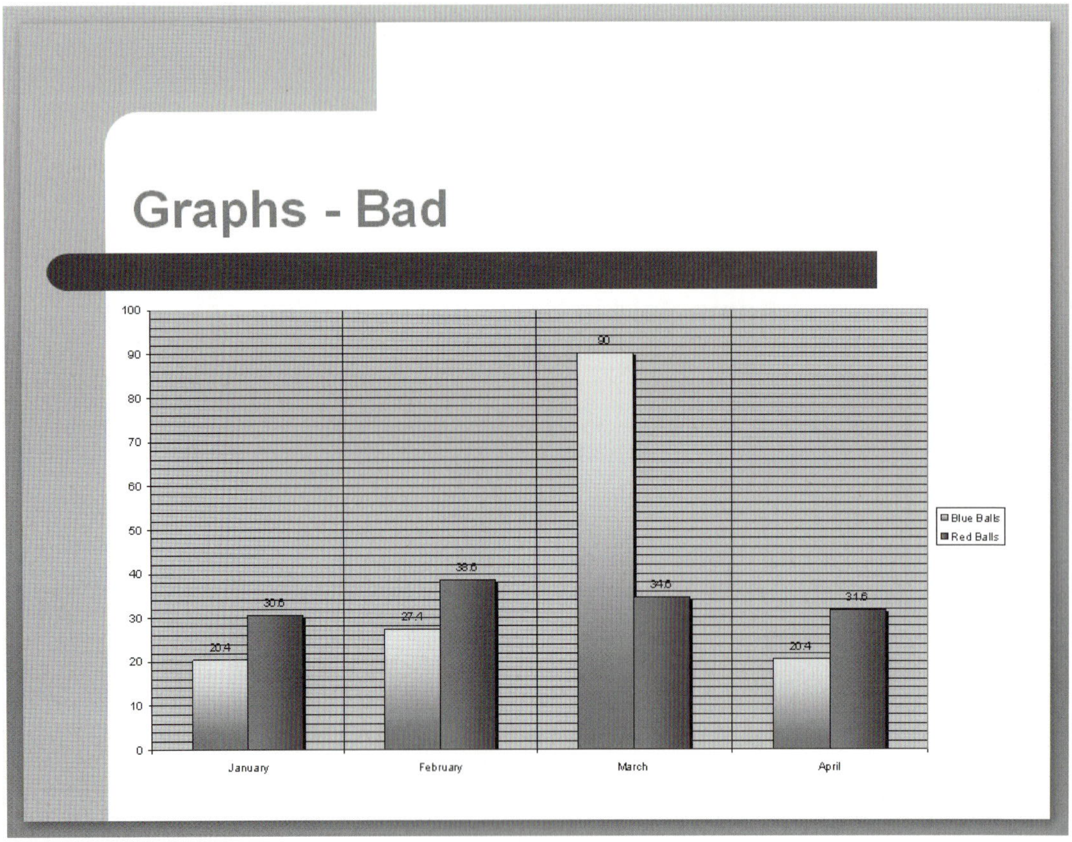

Comments

- 그래픽 개체의 배경 또한 간단하게 표현하는 것이 효과적이다.
- 그래픽 개체의 특징은 다음과 같다.
- 간결하고 명료해서 한 눈에 알 수 있어야 한다.
- 체계적이고 작성자의 의도를 알 수 있어야 하며 본문과 독립적인 개체이나 내용과는 상관관계가 깊어야 한다.

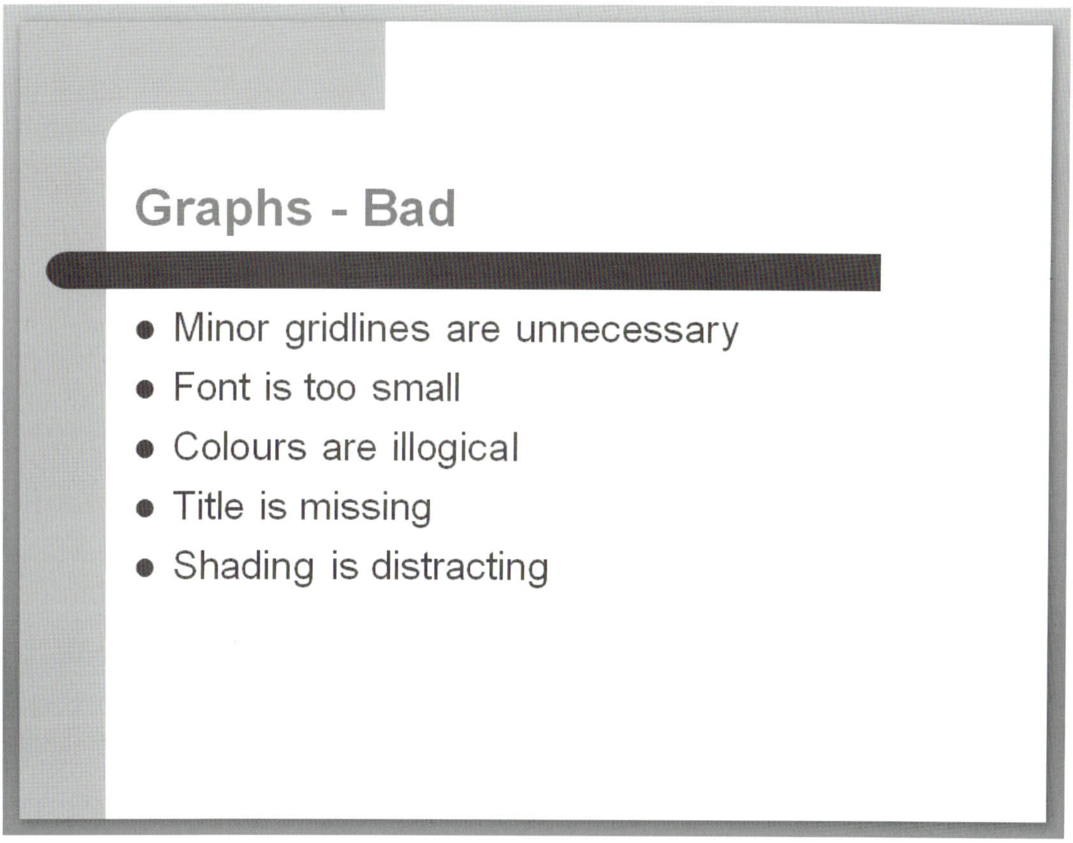

Comments

- 장식 효과나 색깔 사용은 전체적인 느낌이나 리듬에 맞는지 확인하고, 어떤 개체를 사용하든 사용 형식에 맞는지 반드시 확인하자.

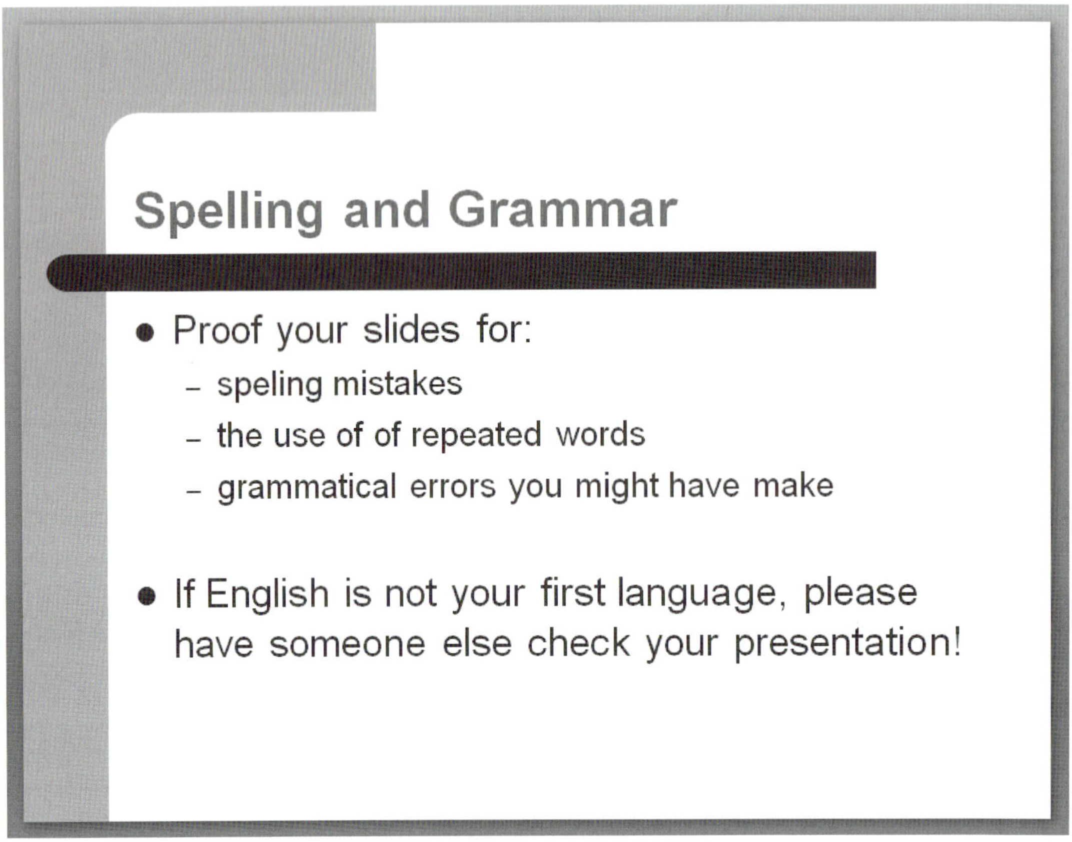

Comments

- 발표자료는 시각 자료이기 때문에 문법오류는 반드시 피해야 한다.
- 영문 자료를 만들 때는 원어민이나 영어를 잘하는 사람에게 검토를 받도록 하자.

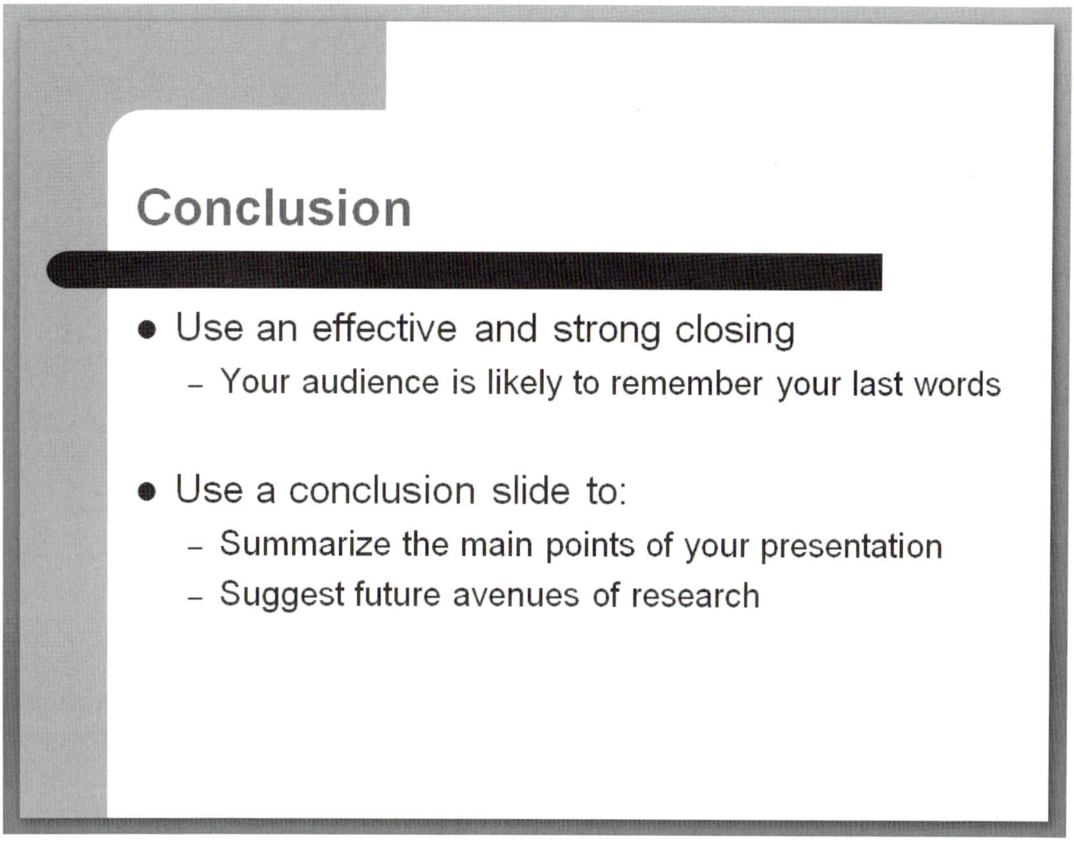

Comments

- 마지막으로 발표자료의 결론 부분을 두어 다시 한번 발표한 내용에 대해 정리를 하는 것이 청중들의 이해를 돕는데 아주 중요하다 하겠다.

References [참고문헌]

[1] 융합 그리고 넥스트, 손병희 저, 교우사, 2015
[2] Cambridge Advanced Learner's Dictionary, Cambridge University Press
[3] Collins COBUILD Intermediate Learner's Dictionary, Heinle & Heinle Publishers
[4] Longman Essential Activator, Longman National Corpus
[5] English Grammar in Use, Raymond Murphy, Cambridge University Press

14과 연습문제

01 주제를 정한뒤 영문 발표자료를 만들어 보자.

연습문제 해답

1장 정답_ 1. C 2. A 3. B 4. E 5. D 6. H 7. F 8. G

Answers for Exercise

1. Ubiquitous computing is not a specific technology, but a scenario in which computers become more numerous and fade into the background, providing information to human users and embedding intelligence and computing capabilities in seemingly everyday objects.
2. Ubiquitous computing–almost imperceptible, but everywhere around us–is rapidly becoming a reality. How will it change us? How can we shape its emergence?
3. Ubiquitous computing (ubicomp) integrates computation into the environment, rather than having computers which are distinct objects. Promoters of this idea hope that embedding computation into the environment and everyday objects would enable people to interact with information–processing devices more naturally and casually than they currently do, and in ways that suit whatever location or context they find themselves in.
4. Ubiquitous computing and wearable computing have been posed as polar opposites even though they are often applied in very similar applications.
5. Embedded and ubiquitous computing is an exciting paradigm that promises to provide computing and communication services to the end users all the time and everywhere. Its systems are now invading in every aspect of our daily life and promise to revolutionize our life much more profoundly than elevators, electric motors or even personal computer evolution ever did. The emergence of this technology is a natural outcome of research and technological advances in a variety of areas including embedded systems, pervasive computing and communications, wireless networks, mobile computing, distributed computing and agent technologies.
6. Ubiquitous computing, or calm technology, is a paradigm shift where technology becomes virtually invisible in our lives. Instead of having a desk–top or lap–top machine, the technology we use will be embedded in our environment.
7. The idea behind ubiquitous computing is to surround ourselves with computers and software that are carefully tuned to offer us unobtrusive assistance as we navigate through our work and personal lives.
8. The any place/any time/any means vision of ubiquitous computing has explosive impact on academics, industry, government and daily life. This emergence is the natural result of research and technological advances in wireless and sensor networks, embedded systems, mobile computing, distributed computing, agent technologies, autonomic computing and communication.
9. Pervasive computing is the trend towards increasingly ubiquitous (another name for the movement is ubiquitous computing), connected computing devices in the environment, a

trend being brought about by a convergence of advanced electronic – and particularly, wireless – technologies and the Internet.

10. At their core, all models of ubiquitous computing (also called pervasive computing) share a vision of small, inexpensive, robust networked processing devices, distributed at all scales throughout everyday life and generally turned to distinctly common-place ends.

2장 정답_ 1. A 2. C 3. B 4. E 5. D 6. F 7. G

Answers for Exercise

1. Computational video includes the study and application of the processing of streamed video data. This field of research is emerging from the convergence of two technologies: digital cameras and high performance computing and high bandwidth networks.
2. Most people appear to think that the future of Digital Media is convergence, that it is inevitable that TVs and Computers are going to converge and become a single device. Some people however think that this will not happen, and they may have good reason for thinking so.
3. Convergence is the coming together of two or more disparate disciplines or technologies. For example, the so-called fax revolution was produced by a convergence of telecommunications technology, optical scanning technology, and printing technology.
4. In graphics, convergence refers to how sharply an individual color pixel on a monitor appears. Each pixel is composed of three dots -- a red, blue, and green one. If the dots are badly misconverged, the pixel will appear blurry. All monitors have some convergence errors, but they differ in degree.
5. A computer convergence system includes a convergence functionality module, a computer, and a display device. The convergence functionality module includes a first input for receiving a first video signal and a second input for receiving a second video signal.
6. With the emergence of the Web as a publishing and programming medium, as well as the availability of high-level development tools, the distinctions between algorithm and presentation are blurring. Today, a programmer needs to understand and apply principles of graphic design in order to develop applications that are usable and attractive to the user. Likewise, a graphic designer developing electronic media must understand and apply programming principles to control the dynamic behavior of the media. Today I will address the convergence of programming and graphic design, from both the perspectives of a computer scientist and a graphic designer.
7. Digital fusion – multimedia* – desktop media – is described as the convergence of computing, television, audio, printing and telecommunications. Each of the components has been significant. Bringing them together results in the whole having greater impact than each individual part and is one of the industry's most significant developments.

8. The convergence of digital technologies and their use will impact the future of teleconferencing, distance education and business.
9. Convergence can be defined as the integration of telephony, computing and media technologies, and hence the integration of the businesses, markets and the social interaction associated with them.
10. Technological convergence is already a common phenomenon in today's ICT devices. Computers are able to function as video and audio players, even having editing capabilities which is traditionally the area of professional broadcasting. The integration of internet connectivity, mobile telephony, computing power, media storage, digital camera, mobile gaming and digital audio player are already apparent in many of today's devices such as mobile phones and personal digital assistants.

3장 정답_ 1. C 2. A 3. B 4. D

Answers for Exercise

1. Ubiquitous computing is not a specific technology, but a scenario in which computers become more numerous and fade into the background, providing information to human users and embedding intelligence and computing capabilities in seemingly everyday objects.
2. Ubiquitous computing-almost imperceptible, but everywhere around us-is rapidly becoming a reality. How will it change us? How can we shape its emergence?
3. Ubiquitous computing (ubicomp) integrates computation into the environment, rather than having computers which are distinct objects. Promoters of this idea hope that embedding computation into the environment and everyday objects would enable people to interact with information-processing devices more naturally and casually than they currently do, and in ways that suit whatever location or context they find themselves in.
4. Ubiquitous computing and wearable computing have been posed as polar opposites even though they are often applied in very similar applications.
5. Embedded and ubiquitous computing is an exciting paradigm that promises to provide computing and communication services to the end users all the time and everywhere. Its systems are now invading in every aspect of our daily life and promise to revolutionize our life much more profoundly than elevators, electric motors or even personal computer evolution ever did. The emergence of this technology is a natural outcome of research and technological advances in a variety of areas including embedded systems, pervasive computing and communications, wireless networks, mobile computing, distributed computing and agent technologies.
6. Ubiquitous computing, or calm technology, is a paradigm shift where technology becomes virtually invisible in our lives. Instead of having a desk-top or lap-top machine, the technology we use will be embedded in our environment.

7. The idea behind ubiquitous computing is to surround ourselves with computers and software that are carefully tuned to offer us unobtrusive assistance as we navigate through our work and personal lives.
8. The any place/any time/any means vision of ubiquitous computing has explosive impact on academics, industry, government and daily life. This emergence is the natural result of research and technological advances in wireless and sensor networks, embedded systems, mobile computing, distributed computing, agent technologies, autonomic computing and communication.
9. Pervasive computing is the trend towards increasingly ubiquitous (another name for the movement is ubiquitous computing), connected computing devices in the environment, a trend being brought about by a convergence of advanced electronic – and particularly, wireless – technologies and the Internet.
10. At their core, all models of ubiquitous computing (also called pervasive computing) share a vision of small, inexpensive, robust networked processing devices, distributed at all scales throughout everyday life and generally turned to distinctly common-place ends.

4장 정답_ 1. D 2. B 3. A 4. C 5. E 6. H 7. G 8. F

Answers for Exercise

1. Radio-frequency identification (RFID) is the use of an object (typically referred to as an RFID tag) applied to or incorporated into a product, animal, or person for the purpose of identification and tracking using radio waves. Some tags can be read from several meters away and beyond the line of sight of the reader.
2. Long checkout lines at the grocery store are one of the biggest complaints about the shopping experience. Soon, these lines could disappear when the ubiquitous Universal Product Code (UPC) bar code is replaced by smart labels, also called radio frequency identification (RFID) tags. RFID tags are intelligent bar codes that can talk to a networked system to track every product that you put in your shopping cart.
3. Motorola's RFID solutions enable you to advance your business to a new level of efficiency by providing greater visibility into your inventory as it moves across the supply chain. With real-time tracking information, you'll always know where your critical business assets are.
4. The federal government now requires US citizens to present identity documents when entering the US. These documents incorporate RFID technology that jeopardizes the privacy and security of US travelers. EPIC has previously urged the State Department to abandon the proposal.

5. Although there are lingering concerns by privacy advocates over the use of RFID, there's no doubt that the use of RFID tags continues to increase. According to a new report from

In-Stat, over 1 billion tags were produced last year, and by 2010, that total will rise to 33 billion. Wal-Mart's requiring of their suppliers to use RFID tags are certainly one reason for the projected growth, as well as potential applications in both the industrial and retail sectors, such as the tracking of inventory.

6. RFID is short for radio frequency identification, a technology similar in theory to bar code identification. With RFID, the electromagnetic or electrostatic coupling in the RF portion of the electromagnetic spectrum is used to transmit signals. An RFID system consists of an antenna and a transceiver, which read the radio frequency and transfer the information to a processing device, and a transponder, or tag, which is an integrated circuit containing the RF circuitry and information to be transmitted.
7. Radio frequency identification (RFID) is a generic term that is used to describe a system that transmits the identity (in the form of a unique serial number) of an object or person wirelessly, using radio waves. It's grouped under the broad category of automatic identification technologies.
8. RFID tags, a technology once limited to tracking cattle, are tracking consumer products worldwide. Many manufacturers use the tags to track the location of each product they make from the time it's made until it's pulled off the shelf and tossed in a shopping cart.
9. RFID is in use all around us. If you have ever chipped your pet with an ID tag, used EZPass through a toll booth, or paid for gas using SpeedPass, you've used RFID. In addition, RFID is increasingly used with biometric technologies for security.
10. Today, RFID is used in enterprise supply chain management to improve the efficiency of inventory tracking and management. However, a fair cost-sharing mechanism between retailer and consumer good suppliers is essential for a sustainable RFID adoption to take place. In addition, it is also necessary for consumer good suppliers to be able to justify their investments in RFID technology with rational motives and justified returns.

5장 정답_ 1. C 2. A 3. D 4. B 5. E 6. G 7. F

Answers for Exercise

1. A wireless sensor network (WSN) consists of spatially distributed autonomous sensors to cooperatively monitor physical or environmental conditions, such as temperature, sound, vibration, pressure, motion or pollutants. The development of wireless sensor networks was motivated by military applications such as battlefield surveillance. They are now used in many industrial and civilian application areas, including industrial process monitoring and control, machine health monitoring, environment and habitat monitoring, healthcare applications, home automation, and traffic control.
2. As wireless sensors come onto the market and become a practical option, sensor networks take on a whole new dimension. The question arises: How can traditional network topologies

be adapted to the new communications medium?

3. According to GainSpan, its technology provides the lowest power consuming Wi-Fi single chip solution for wireless sensor networks and other embedded applications, allowing devices to run for up to 10 years on a single AA battery.
4. WSN, otherwise known as Ubiquitous Sensor Networks (USN) will be used to automatically monitor building occupancy and climate control, oil and gas pumping, forest fires, avalanches, the effect and likely effect of hurricanes, traffic, hospitals and much more over wide areas over the coming years. The WSN industry will grow rapidly to exceed two billion dollars in yearly sales in the next ten years and about six billion dollars during the following ten years. In addition, new this year is coverage of energy harvesting and energy storage for small electronic devices.
5. A typical sensor network device comprises the following components some of which are optional: power supply, microcontroller, wireless communication, wired communication, sensor, local storage, and real time clock systems. The principal idea is that the sensors are connected to a tiny computer that coordinates the measurement, preprocesses, stores and delivers the information.
6. Researchers have long been promising that wireless sensor networks will bring a new wave of smart appliances that can communicate with each other, and with central servers when they need attention. The technology for these appliances has existed for some time now. However, the commercial availability of such products has been slow to market for one primary reason – the lack of structured business model.
7. Who said Wireless Sensor Networks had to be small personal area networks? According to Network World, researchers at Penn State University are developing a sensor that would be attached to cars that would allow it to talk to other cars, forming a large area sensor network.
8. It was predicted back in 2005 that WSN would be a disruptive technology that changed the way we viewed the world. Today, market research still predicts that the WSN revolution is "just around the corner".
9. WSN is a great technology, but that's just it. Its another technology. The technology alone doesn't do anything; it needs to be applied. And for most of the things that it can be applied to, there is often a cheaper, lower-tech, more reliable method.
10. As wireless sensors become real commodities on the market, new options or new arguments for old options are causing professionals to consider network strategies once ruled out. Let's look at the three classic network topologies (point-to-point, multidrop, and web), assess their strengths and weaknesses, and look at how the rules have changed now that wireless systems are coming online.

6장 정답_ 1. A 2. D 3. B 4. C 5. F 6. E

Answers for Exercise

1. A mobile ad hoc network (MANET), sometimes called a mobile mesh network, is a self-configuring network of mobile devices connected by wireless links.
2. So what is a mesh network? Think of the electronic equivalent of party guests passing on the message that a child is missing until that child is found – many short range communications, in many directions – like a mesh – add up to finding the child further away. All those people were able to pass on a message in an ad hoc network that dissolved when its task is done. In the electronic equivalent, WSN can form ad hoc networks where many small ranges add up to a big one thus permitting the tags to be much smaller, lower cost and more reliable than would otherwise be the case. However, here the network operates without human intervention. It performs its tasks silently and automatically.
3. An ad-hoc network is a local area network (LAN) that is built spontaneously as devices connect. Instead of relying on a base station to coordinate the flow of messages to each node in the network, the individual network nodes forward packets to and from each other. In Latin, ad hoc literally means "for this," meaning "for this special purpose" and also, by extension, improvised or impromptu.
4. On wireless computer networks, ad-hoc mode is a method for wireless devices to directly communicate with each other. Operating in ad-hoc mode allows all wireless devices within range of each other to discover and communicate in peer-to-peer fashion without involving central access points (including those built in to broadband wireless routers).
5. Ad hoc networking might seem like a much smarter alternative to using a wireless router, but it does have a couple of disadvantages: If the computer connected to the Internet is shut down, all computers that are part of the ad hoc network lose their Internet access. To connect to the Internet, one computer always needs a wired network connection.
6. On networks with wireless routers, the router has the important job of forwarding communications from the computers on your home network to the Internet. On ad hoc networks, you must designate one computer to serve this role. The computer you choose must have a wired connection to the Internet, and it should be left on whenever you want to be able to use your other computers.
7. I've received many inquiries from Windows XP residential users asking how to wirelessly network a second or third computer without the expense of a wireless router or a router and a stand-alone wireless access point (WAP). This is sometimes called an ad hoc wireless network. Building an ad hoc 802.11b wireless network using the graphical user interface in Windows XP takes just a few minutes. Because it's powered by wireless zero configuration, you won't need to run a single cable or hunt for an electrical outlet! Add Internet Connection Sharing on the host computer and all connected computers will be surfing the net wirelessly in no time at all.
8. Ad Hoc Networks are self organizing, self healing, distributed networks which most often employ wireless transmission techniques.
9. Ad hoc networks, which cover a variety of network paradigms for specific purposes, such as mobile ad hoc networks, sensor networks, vehicular networks, underwater networks, underground networks, local area networks, personal area networks, home networks, etc., promise a broad range of applications in civilian, commercial, and military areas.
10. Smart Ad Hoc Networks are capable of analyzing the radio propagation environment they operate in to optimize performance. This typically requires that the network

nodes have positioning capability as well as memory to recall geographically local condition.

7장 정답_ 1. C 2. B 3. A 4. D 5. E

Answers for Exercise

1. Telematics is the convergence of telecommunications and information processing, the term later evolved to refer to automation in automobiles, such as the invention of the emergency warning system for vehicles. GPS navigation, integrated hands-free cell phones, wireless safety communications and automatic driving assistance systems all are covered under the telematics umbrella.
2. Telematics technologies are self-orientating open network architecture structure of variable programmable intelligent beacons developed for application in the development of intelligent vehicles — with target intent to accord (blend, or mesh) warning information with surrounding vehicles in the vicinity of travel, intra-vehicle, and infrastructure. Emergency warning system for vehicles telematics particularly developed for international harmonization and standardization of vehicle-to-vehicle — infrastructure-to-vehicle — and vehicle-to-infrastructure real-time Dedicated Short Range Communication (DSRC) systems.
3. Telematics is the blending of computers and wireless telecommunications technologies, ostensibly with the goal of efficiently conveying information over vast networks to improve a host of business functions or government-related public services. The most notable example of telematics may be the Internet itself, since it depends on a number of computer networks connected globally through telecommunication backbones.
4. Cold store freight trailers that are used to deliver fresh or frozen foods are increasingly incorporating telematics to gather time-series data on the temperature inside the cargo container, both to trigger alarms and record an audit trail for business purposes. An increasingly sophisticated array of sensors, many incorporating RFID technology, are being used to ensure that temperature throughout the cargo remains within food-safety parameters.
5. Major automakers are equipping new prototype vehicles with wireless-based services controlled by voice commands. This kind of telematics could enable motorists to perform a variety of wireless functions such as accessing the Internet, receiving or sending e-mail, downloading digital audio and video files, or obtaining "smart" transportation information.
6. Wireless vehicle safety communications telematics aid in car safety and road safety. It is an electronic sub-system in a car or other vehicle for the purpose of exchanging safety information, about such things as road hazards and the locations and speeds of vehicles, over short range radio links. This may involve temporary ad hoc wireless local area networks.

7. The telematics industry is not limited to automotive applications. Other applications are being studied or developed for monitoring water and air pollution, for medical informatics and health care, and for distance learning. Many European countries are developing uniform policies to integrate telematics applications into government, business and education.
8. Telematics is now a somewhat dated term, but has been used to describe the convergent action of computer and communication (terrestrial and satellite) technologies. Another related term is ICT – Information and Communications Technologies – essentially the same thing. In other words, telematics is about the merging of computers and telecommunications. Telematic technologies include satellite television, the use of computers to mediate communication known as CMC, visual collaboration through the use of video conferencing and of course the Internet.
9. With the current financial constraints and the increasing pressure on education providers to perform better, telematics offers a viable solution for sustainability and growth. Widening catchment areas and increasing student numbers without direct access to new resources such as more classrooms and staff can be achieved using telematics. There is also the move towards lifelong learning to consider.
10. Telematics comprise electronic, electromechanical, and electromagnetic devices — usually silicon micromachined components operating in conjunction with computer controlled devices and radio transceivers to provide precision repeatability functions (such as in robotics artificial intelligence systems) emergency warning validation performance reconstruction.

8장 정답_ 1. D 2. C 3. A 4. B 5. F 6. E

Answers for Exercise

1. An embedded system is a computer system designed to perform one or a few dedicated functions, often with real-time computing constraints. It is embedded as part of a complete device often including hardware and mechanical parts. In contrast, a general-purpose computer, such as a personal computer, is designed to be flexible and to meet a wide range of an end-user's needs. Embedded systems control many of the common devices in use today.
2. The challenge of programming an embedded system is that it is difficult to get real-time feedback from the system without a display. It is common to use a simple serial interface for diagnostic purposes, for example by connecting to a PC running terminal software via a RS-232 to USB adapter. Also, embedded systems usually have very strict memory limitations, processor limitations, and speed limitations that must play a factor in designing an embedded system, and programming an embedded computer.
3. Physically, embedded systems range from portable devices such as digital watches and MP3 players, to large stationary installations like traffic lights, factory controllers, or the systems

controlling nuclear power plants. Complexity varies from low, with a single microcontroller chip, to very high with multiple units, peripherals and networks mounted inside a large chassis or enclosure.

4. Embedded systems evolved with general-purpose computer systems. They were at the forefront of the use and justification of computer products, the closest to the field. This is where the theories are making or breaking.
5. In today's world, embedded systems are everywhere -- homes, offices, cars, factories, hospitals, plans and consumer electronics. Their huge numbers and new complexity call for a new design approach, one that emphasizes high-level tools and hardware/software tradeoffs, rather than low-level assembly-language programming and logic design.
6. An embedded system is a specialized computer system that is part of a larger system or machine. Typically, an embedded system is housed on a single microprocessor board with the programs stored in ROM. Virtually all appliances that have a digital interface -- watches, microwaves, VCRs, cars -- utilize embedded systems. Some embedded systems include an operating system, but many are so specialized that the entire logic can be implemented as a single program.
7. In the past, most of the embedded research and development was identified with real-time systems and industrial settings, but things have been changing since. With wider deployment of computers, the need for embedded systems has increased.
8. In general, "embedded system" is not an exactly defined term, as many systems have some element of programmability. For example, handheld computers share some elements with embedded systems such as the operating systems and microprocessors which power them, but are not truly embedded systems, because they allow different applications to be loaded and peripherals to be connected.
9. An embedded system is some combination of computer hardware and software, either fixed in capability or programmable, that is specifically designed for a particular function. Industrial machines, automobiles, medical equipment, cameras, household appliances, airplanes, vending machines and toys (as well as the more obvious cellular phone and PDA) are among the myriad possible hosts of an embedded system.
10. Over recent years, embedded systems have gained an enormous amount of processing power and functionality. Many of the formerly external components can now be integrated into a single System-on-Chip. This tendency has resulted in a dramatic reduction in the size and cost of embedded systems. As a unique technology, the design of embedded systems is an essential element of many innovations.

9장 정답_ 1. C 2. A 3. B 4. D 5. F 6. E

Answers for Exercise

1. A robot is a virtual or mechanical artificial agent. In practice, it is usually an

electro-mechanical machine which is guided by computer or electronic programming, and is thus able to do tasks on its own. Another common characteristic is that by its appearance or movements, a robot often conveys a sense that it has intent or agency of its own.

2. Typical applications of robots include welding, painting, assembly, pick and place, packaging and palletizing, product inspection, and testing, all accomplished with high endurance, speed, and precision.
3. Robots have captured the imagination of inventors and storytellers for centuries. While the birth of automation technology can be traced to the Greeks, the first industrial robots were not produced until the 1970's. Today's cutting edge humanoid robots are the result of years of trial and error.
4. Some robots are programmed to faithfully carry out specific actions over and over again (repetitive actions) without variation and with a high degree of accuracy. These actions are determined by programmed routines that specify the direction, acceleration, velocity, deceleration, and distance of a series of coordinated motions.
5. Perhaps the most popular applications of robots is in industrial welding. The repeatability, uniformity quality, and speed of robotic welding is unmatched. The two basic types of welding are spot welding and arc welding, although laser welding is done. Some environmental requirements should be considered for a successful operation.
6. Who invented the term robots? In 1956, George Devil and Joseph Engelberger formed the world's first robot company, but writers have been dreaming about robots long before that.
7. An industrial robot is officially defined by ISO as an automatically controlled, reprogrammable, multipurpose manipulator programmable in three or more axes. The field of robotics may be more practically defined as the study, design and use of robot systems for manufacturing (a top-level definition relying on the prior definition of robot).
8. Modern industrial robots are true marvels of engineering. A robot the size of a person can easily carry a load over one hundred pounds and move it very quickly with a repeatability of +/-0.006 inches. Furthermore these robots can do that 24 hours a day for years on end with no failures whatsoever. Though they are reprogrammable, in many applications (particularly those in the auto industry) they are programmed once and then repeat that exact same task for years.
9. Robots are used in a wide range of industrial applications. The earliest applications were in materials handling, spot welding, and spray painting. Robots were initially applied to jobs that were hot, heavy, and hazardous such as die casting, forging, and spot welding.
10. Interest in robotics increased in the late 1970s and many US companies entered the field, including large firms like General Electric, and General Motors (which formed joint venture FANUC Robotics with FANUC LTD of Japan). U.S. startup companies included Automatix and Adept Technology, Inc. At the height of the robot boom in 1984, Unimation was acquired by Westinghouse Electric Corporation for 107 million U.S. dollars. Westinghouse sold Unimation to Stäubli Faverges SCA of France in 1988, which is still making articulated robots for general industrial and cleanroom applications and even bought the robotic division of Bosch in late 2004.

10장 정답_ 1. D 2. A 3. E 4. B 5. F 6. C

Answers for Exercise

1. Digital television (DTV) is the sending and receiving of moving images and sound by discrete (digital) signals, in contrast to the analog signals used by analog TV.
2. Your DTV reception can be affected by terrain, trees, buildings, the weather, damaged equipment, as well as antenna type, location, and orientation. It can be improved just by changing the location of the antenna you're using now. Moving your antenna away from other objects and structures, or placing it higher, can often improve reception. The performance of outdoor antennas can degrade over time due to exposure to the weather. Also, you may consider installing a signal booster, which should improve reception.
3. Digital television supports many different picture formats defined by the combination of size, aspect ratio (height to width ratio) and interlacing. With terrestrial broadcasting in the USA, the range of formats can be coarsely divided into two categories: HDTV and SDTV.
4. Digital television (DTV) is an innovative type of over-the-air broadcasting technology that enables TV stations to provide dramatically clearer pictures and better sound quality.
5. Digital TV commenced in 2001 in metropolitan areas and digital signals have been broadcast in many regional areas since then. Digital broadcasts are currently transmitted alongside analog broadcasts, but when the analog signals are switched off, only TVs which are capable of receiving digital signals through a set top box or built-in tuner will work.
6. Digital television, or DTV, is a technology that gives viewers crystal-clear pictures and sound, and more programming choices than ever before.
7. Your DTV reception can be affected by nearby moving vehicles, such as cars, trucks, trains and airplanes. In some instances, shadowing or reflections from these vehicles may cause your digital picture to temporarily break-up or even disappear completely. If this occurs, you should try moving or reorienting your antenna to find a position that provides the most reliable reception. If you are using an indoor antenna, switching to an outdoor antenna system which may include a directional antenna or rotor could improve reception. In severe cases it may not be possible to completely eliminate the effect of nearby traffic. If reception remains unsatisfactory due to these disruptions, viewers may wish to consider alternatives such as cable or satellite service.
8. High-definition television (HDTV), one of several different formats that can be transmitted over DTV, uses one of two formats: 1280 × 720 pixels in progressive scan mode (abbreviated 720p) or 1920 × 1080 pixels in interlace mode (1080i). Each of these utilizes a 16:9 aspect ratio. (Some televisions are capable of receiving an HD resolution of 1920 × 1080 at a 60 Hz progressive scan frame rate — known as 1080p60, but this standard is not currently used for transmission.) HDTV cannot be transmitted over current analog channels.
9. DTV is more efficient and more flexible than the traditional way of broadcasting known as analog. For example, DTV makes it possible for stations to broadcast multiple channels of free programming all at once (called multicasting), instead of broadcasting one channel

at a time. DTV technology can also be used to deliver future interactive video and data services that analog technology can't provide.

10. There are a number of different ways to receive digital television. One of the oldest means of receiving DTV (and TV in general) is using an antenna (known as an aerial in some countries). This way is known as Digital Terrestrial Television (DTT). With DTT, viewers are limited to whatever channels the antenna picks up. Signal quality will also vary.

11장 정답_ 1. B 2. A 3. C

Answers for Exercise

1. A content management system (CMS) such as a document management system (DMS) is a computer application used to manage work flow needed to collaboratively create, edit, review, index, search, publish and archive various kinds of digital media and electronic text.
2. Content Management Systems come in all shapes and sizes. Chances are, if you've put up a Web page on a free Web page server or ISP you've had access to a rudimentary content management system.
3. Oracle Universal Content Management (UCM) is the industry's most unified enterprise content management platform that enables you to leverage market-leading document management, Web content management, digital asset management, and records retention functionality to build and complement your business applications. Building a strategic enterprise content management infrastructure for content and applications helps you to reduce costs, easily share content across the enterprise, minimize risk, automate expensive, time-intensive and manual processes, and consolidate multiple Web sites onto a single platform for centralized management. Through user-friendly interfaces, roles-based authentication and security models, Oracle Universal Content Management empowers users throughout the enterprise to view, collaborate on or retire content, ensuring that all accessible distributed or published information is secure, accurate and up-to-date.
4. There are two fundamental elements to content management: (1) storing stuff in a content repository, and (2) supporting the workflow of a group of people engaged in putting stuff into that repository.
5. CMSs are frequently used for storing, controlling, versioning, and publishing industry-specific documentation such as news articles, operators' manuals, technical manuals, sales guides, and marketing brochures. The content managed may include computer files, image media, audio files, video files, electronic documents, and Web content. These concepts represent integrated and interdependent layers. There are various nomenclatures known in this area: Web Content Management, Digital Asset Management, Digital Records Management, Electronic Content Management and so on. The bottom line

for these systems is managing content and publishing, with a workflow if required.

6. Effective content management systems support an organization's business processes for acquiring, filtering, organizing, and controlling access to information. Because no two organizations use identical business processes, content management systems in practice can be compared to snowflakes -- no two such systems will look or behave exactly alike!
7. YouTube has created an advanced set of copyright policies and content management tools to give rights holders control of their content. YouTube provides content management solutions for rights holders of all sizes across the world, and provides tools to cater to the specific needs of various rights owners.
8. Content management, or CM, is the set of processes and technologies that support the collection, managing, and publishing of information in any form or medium. In recent times this information is typically referred to as content or, to be precise, digital content. Digital content may take the form of text, such as documents, multimedia files, such as audio or video files, or any other file type which follows a content lifecycle which requires management.
9. There are plenty of options when it comes to picking a content management system for a development project. Depending on how advanced you need the CMS to be, what language it's built in, and who is going to be using it, it can be a nightmare trying to find the "perfect" CMS for a project.
10. Content management practices and goals vary with mission. News organizations, e-commerce websites, and educational institutions all use content management, but in different ways. This leads to differences in terminology and in the names and number of steps in the process.

12장 정답_ 1. B 2. C 3. D 4. E 5. A

Answers for Exercise

1. Whether your computer runs Microsoft® Windows®, Apple's Mac OS, LINUX, or something else, the issues are the same and will remain so as new versions of your system are released. The key is to understand the security-related problems that you need to think about and solve.
2. Computer security is a branch of technology known as information security as applied to computers and networks. The objective of computer security includes protection of information and property from theft, corruption, or natural disaster, while allowing the information and property to remain accessible and productive to its intended users. The terms computer system security, means the collective processes and mechanisms by which sensitive and valuable information and services are protected from publication, tampering or collapse by unauthorized activities or untrustworthy individuals and unplanned events respectively.

3. No matter how a home computer is connected to the Internet, intruders' attacks are often successful. Many home computer owners don't realize that they need to pay attention to computer security. In the same way that you are responsible for having insurance when you drive a car, you need to also be responsible for your home computer's security. This document explains how some parts of the Internet work and then describes tasks you can do to improve the security of your home computer system. The goal is to keep intruders and their programs off your computer.
4. The term computer security is used frequently, but the content of a computer is vulnerable to few risks unless the computer is connected to other computers on a network. As the use of computer networks, especially the Internet, has become pervasive, the concept of computer security has expanded to denote issues pertaining to the networked use of computers and their resources.
5. In an email message, an intruder can easily fabricate where the email came from. But this information forging - called spoofing by intruders and security professionals - is not limited to just email. In fact, the basic unit of information transferred on the Internet - called a packet - can also be easily forged or spoofed.
6. Think of your computer as you would your house, your apartment, or your condo. What do you know about how that living space works, what do you routinely do to keep it secure, and what have you installed to improve its security?
7. The technologies of computer security are based on logic. As security is not necessarily the primary goal of most computer applications, designing a program with security in mind often imposes restrictions on that program's behavior.
8. While confidentiality, integrity, and authenticity are the most important concerns of a computer security manager, privacy is perhaps the most important aspect of computer security for everyday Internet users. Although users may feel that they have nothing to hide when they are registering with an Internet site or service, privacy on the Internet is about protecting one's personal information, even if the information does not seem sensitive. Because of the ease with which information in electronic format can be shared among companies, and because small pieces of related information from different sources can be easily linked together to form a composite of, for example, a person's information seeking habits, it is now very important that individuals are able to maintain control over what information is collected about them, how it is used, who may use it, and what purpose it is used for.
9. Information security policies underpin the security and well being of information resources. They are the foundation, the bottom line, of information security within an organization.
10. There are various strategies and techniques used to design security systems. However there are few, if any, effective strategies to enhance security after design. One technique enforces the principle of least privilege to great extent, where an entity has only the privileges that are needed for its function. That way even if an attacker gains access to one part of the system, fine-grained security ensures that it is just as difficult for them to access the rest.

13장 정답_

1. 다음을 영작하여 표현해 보시오.
   - I'm happy to introduce our new line of products today.
   - Before I wrap up my talk, are there any questions about my presentation?

2. 다음 그림에서 그래프의 변화를 표현해 보시오.

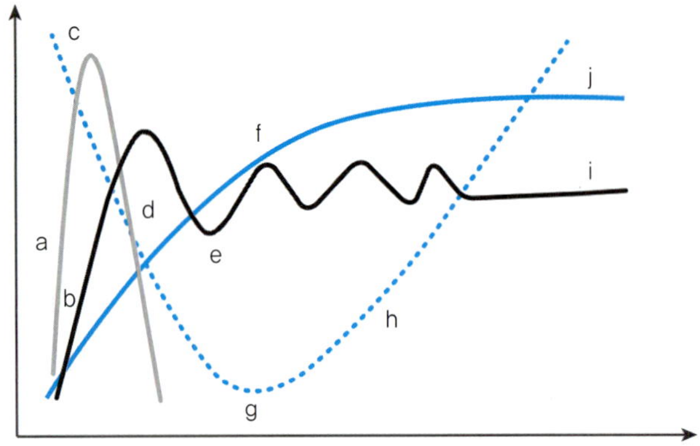

표현 어구를 골라본다.

- decrease, fall, level out(off), remain constant, touch the bottom, increase to, rise ascend, grow steadily, rise dramatically, drop suddenly, reach a peak, converge

색 인

손병희

연세대학교 전기전자공학부에서 석, 박사를 마쳤으며, 현재 인하공업전문대학 컴퓨터시스템과 교수로 재직중이다. 관심 분야는 감성을 이용한 홈 네트워킹 시스템과 지능형 임베디드 시스템 등 상황인지 기술이다.

강효선

숭실대학교 졸업 후 홍익대학교 교육대학원에서 영어교육 석사를 마쳤다.
현재 경기도에 소재한 공립학교에서 영어교사로서 중고등학생을 가르치고 있다.

공대생을 위한 전공영어

2015년 08월 25일 초판 인쇄
2015년 09월 03일 초판 발행

저 자 • 손병희 · 강효선
발행자 • 오판근
발행처 • 교우사

주 소 • 서울특별시 동대문구 약령시로 8 교우빌딩 2층
전 화 • 925-2861(代), 편집부 / 925-2825, 홈쇼핑 / 969-5762
팩 스 • 925-2860

이메일 / kyowoo@kyowoo.co.kr
홈페이지 / http://www.kyowoo.co.kr
북스토어(쇼핑몰) / http://www.kyowoo.co.kr

등록번호 / 1994년 3월 24일 제6-0256호

ISBN 979-11-251-0125-3(93740)

가격 18,000원

© KyoWooSa, 2015.
※ 이 책의 무단전재, 복제를 금합니다.